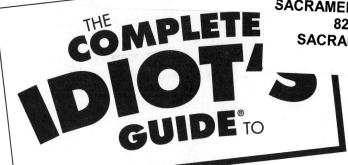

Global Warming

Second Edition

by Michael Tennesen

ALPHA

A member of Penguin Group (USA) Inc.

I dedicate this book to my wife, Maggie, and my mom, Annabelle. I couldn't have done it without you two.

ALPHA BOOKS

Published by the Penguin Group

Penguin Group (USA) Inc., 375 Hudson Street, New York, New York 10014, USA

Penguin Group (Canada), 90 Eglinton Avenue East, Suite 700, Toronto, Ontario M4P 2Y3, Canada (a division of Pearson Penguin Canada Inc.)

Penguin Books Ltd., 80 Strand, London WC2R 0RL, England

Penguin Ireland, 25 St. Stephen's Green, Dublin 2, Ireland (a division of Penguin Books Ltd.)

Penguin Group (Australia), 250 Camberwell Road, Camberwell, Victoria 3124, Australia (a division of Pearson Australia Group Pty. Ltd.)

Penguin Books India Pvt. Ltd., 11 Community Centre, Panchsheel Park, New Delhi—110 017, India

Penguin Group (NZ), 67 Apollo Drive, Rosedale, North Shore, Auckland 1311, New Zealand (a division of Pearson New Zealand Ltd.)

Penguin Books (South Africa) (Pty.) Ltd., 24 Sturdee Avenue, Rosebank, Johannesburg 2196, South Africa

Penguin Books Ltd., Registered Offices: 80 Strand, London WC2R 0RL, England

International Standard Book Number: 978-1-59257-717-0
Library of Congress Catalog Card Number: 2007935860

10 09 08   8 7 6 5 4 3 2 1

Interpretation of the printing code: The rightmost number of the first series of numbers is the year of the book's printing; the rightmost number of the second series of numbers is the number of the book's printing. For example, a printing code of 08-1 shows that the first printing occurred in 2008.

Printed in the United States of America

Note: This publication contains the opinions and ideas of its author. It is intended to provide helpful and informative material on the subject matter covered. It is sold with the understanding that the author and publisher are not engaged in rendering professional services in the book. If the reader requires personal assistance or advice, a competent professional should be consulted.

The author and publisher specifically disclaim any responsibility for any liability, loss, or risk, personal or otherwise, which is incurred as a consequence, directly or indirectly, of the use and application of any of the contents of this book.

Most Alpha books are available at special quantity discounts for bulk purchases for sales promotions, premiums, fundraising, or educational use. Special books, or book excerpts, can also be created to fit specific needs.

For details, write: Special Markets, Alpha Books, 375 Hudson Street, New York, NY 10014.

Publisher: *Marie Butler-Knight*
Editorial Director: *Mike Sanders*
Managing Editor: *Billy Fields*
Acquisitions Editor: *Tom Stevens*
Senior Development Editor: *Phil Kitchel*
Senior Production Editor: *Janette Lynn*

Copy Editor: *Lisanne V. Jensen*
Cartoonist: *Richard King*
Cover Designer: *Kurt Owens*
Book Designer: *Trina Wurst*
Indexer: *Tonya Heard*
Layout: *Chad Dressler*
Proofreader: *John Etchison*

Contents at a Glance

Contents

Foreword

Let's be honest: we're all idiots sometimes. You cruise through life minding your own business and suddenly everyone's talking about tofu, or iPhones, or whether the napkin goes at the top of the place setting or tucked neatly under the fork on the left.

The topic that everyone's talking about these days is global warming. Is it "the greatest hoax ever perpetrated on the American people," as U.S. Senator James Inhofe said a few years ago? Or are we instead "facing a global climate crisis," as former vice president and Nobel laureate Al Gore opined in a recent speech? This book will help you cut through such rhetoric using simple language and a clear treatment of the science.

Whatever you've heard, global warming isn't a hoax. Scientists have long known how greenhouse gases work. In 1827, physicist Jean Baptiste Fourier showed how carbon dioxide, methane, and other greenhouse gases trapped radiation to warm the earth. By the 1890s, chemist Svante Arrhenius (another Nobel laureate) described how emissions from burning coal, oil, and other fossil fuels could warm the earth further.

Today, greenhouse gas concentrations are higher than at any time in the last million years, probably tens of millions. Atmospheric carbon dioxide has increased by a third since the start of the Industrial Revolution, and methane concentrations have more than doubled. How do we know this with certainty? Read Michael Tennesen's book and find out.

We also know that the earth is warming quickly. We know this from weather stations, aircraft measurements, balloons, buoys, and boreholes. Pictures help, too. Satellite images show us that Arctic sea ice in 2007 shrank by 40 percent compared to a typical summer in the 1970s. Fifty years from now, you may be able to canoe to the north pole.

Couldn't natural causes, such as volcanoes or sunspots, be the culprits of warming? No, but don't take my word for it. Read the consensus statements from the U.S. National Academy of Sciences or the Intergovernmental Panel on Climate Change. Read the official statement of the 11,000-member American Meteorological Society: "Because human activities are contributing to climate change, we have a collective responsibility to develop and undertake carefully considered response actions." Not *if* human activities are contributing to climate change, but *because* they are.

If global warming isn't a hoax, is it the "global climate crisis" that Al Gore refers to? Not yet, though if we don't act quickly it will be. This book is helpful because it also presents some practical solutions, such as like improving gas mileage in the cars we drive, or finding more abundant and cheaper sources of renewable energy. Did you realize that Henry Ford's Model T got the same mileage as the U.S. fleet does today—a paltry 20 miles to the gallon? Or that Denmark now generates one fifth of all its electricity using wind power?

Tackling global warming makes environmental and economic sense. This book will show you why, while acknowledging some of the uncertainties that remain.

As the popular comic-strip character Dilbert likes to say, "People are idiots." That observation may be true, but some of us are smart enough to know when to get help. Read all about global warming and make yourself one of the smartest idiots around.

—Rob Jackson

Rob Jackson is the Nicholas Professor of Global Environmental Change at Duke University and founder and co-director of the Climate Change Policy Partnership. He has written several books about the environment, including *The Earth Remains Forever*, and a book of children's poetry, *Animal Mischief*.

Introduction

A friend of mine called the other day. Like most of us, my friend is a concerned citizen. But he didn't want any long explanations. His questions were pointed: What is global warming? What is it going to do? Is it going to affect me in my lifetime?

I tried to be succinct. Global warming is the effect that the greenhouse gases (carbon dioxide, methane, and others) are having on the earth's climate. Although they constitute less than 1 percent of the atmosphere, the greenhouse gases—especially carbon dioxide—act like a blanket covering the globe. Without the greenhouse effect, the average global temperature would be around 0°F (–18°C). With the greenhouse effect, the temperature averages a toasty 59°F (15°C).

The trouble is, the volume of greenhouse gases in the atmosphere has been increasing. This increase is due to the pollution caused by the burning of fossil fuels (coal, oil, and natural gas). It's also caused by the fact that we're cutting down the rainforests, which used to suck up a lot of that carbon dioxide. At the current rate of increase, the volume of greenhouse gases in the atmosphere could more than double by the end of the century.

Most scientists agree that the increase of greenhouse gases in the atmosphere is changing the climate. The controversy is over the amount of change, what that change will bring, and whether it's worth the high repair bill.

The Intergovernmental Panel on Climate Change, a joint project of the World Meteorological Organization and the United Nations Environment Program, estimates that if carbon dioxide in the atmosphere continues to rise, the world will warm up by 2.0 to 11.5°F (1.1 to 6.4°C) by the end of the twenty-first century.

If the temperature rises only 2.0°F (1.1°C), we might be able to live with it. If it rises 11.5°F (6.4°C), there are going to be major problems. An increase in the global average surface temperature of 11.5°F (6.4°C) is as great a change as the one that ended the last ice age—but instead of happening over about 50 centuries, it will happen in less than one. The increase could cause major floods in some parts of the country, droughts in other areas, drown large sections of Florida, and wreak havoc with nature everywhere.

Some results are expected in the next 30 years. Some are happening right now. In the last century, there has been a widespread retreat of mountain glaciers outside the poles. Half the ice in the European Alps has melted. Summer Arctic sea ice is 40 percent thinner. Glacier National Park has already lost more than two thirds of its glaciers. In another 30 years, scientists estimate that Glacier National Park will be glacierless. Four Antarctic ice shelves have collapsed. In 2002, the Larsen-B ice shelf began to disintegrate, and within 35 days, an area that measured 150 miles long by 30 miles wide—more than twice the area of Rhode Island—completely broke up.

This new edition of the book attempts to give you a grasp of the enormity of these issues, the politics behind change, and what you can do about it.

What You'll Learn in This Book

The *Complete Idiot's Guide to Global Warming, Second Edition*, is divided into six parts:

Part 1, "Global Warming Signs," goes into the basics of global warming. I start with predictions of what could happen if we forget about it—just to keep you awake. Next, I discuss the atmospheric chemicals that make up the greenhouse effect. We look at how scientists are studying the history of our climate by looking at ocean bottoms, tree rings, and polar ice. Then I'll take you on a four-billion-year tour of the atmosphere. I'll close with a look at the ozone hole, our first climate catastrophe.

Part 2, "A Historical Perspective," shows you how the climate has made changes—*big* changes—in the past and how we might not be ready for the future. I'll talk about how the climate affected the Vikings and the Mayans, how industrial pollution has tainted the air in the twentieth century, and how cutting down the forests has changed the atmosphere we breathe.

Part 3, "A Meteorological Primer," explains some of the basics of weather and climate. I'll tell you how winds are formed, how rain is made, and how hurricanes are born. I'll talk about El Niño—a sample of what a little warming can do. I'll look at the atmospheric variables that could be causing global warming. I'll also take a look at polar ice, deep ocean currents, and some of the other stuff that could make global warming a lot more dangerous.

Part 4, "The Crystal Ball," discusses some of the effects that global warming might have on Africa, Asia, Europe, Latin America, and our low-lying island nations. I'll also gaze into the future of North America. How will global warming affect commerce, agriculture, and health? You'll learn what might happen to our wildlife, plants, and national parks. And I'll take a look at climate models and how they come up with their predictions for the future.

Part 5, "The Politics," looks at the Kyoto Protocol, how it was built, what it required, and where it stands now. We'll discuss whether the causes of pollution reduction, energy conservation, and rainforest preservation aren't worthwhile all by themselves. I'll look at the international debate, who's skeptical, and who's not. This part also covers how U.S. politics has evolved on the question. We'll look at how the world got together to solve the Antarctic ozone hole. And finally, I'll sum up the argument.

Part 6, "The Solution," goes beyond the controversy into the answer. What can you do around the home and in the garden to reduce energy consumption? What about that car? Are SUVs and cheap gas such a great deal after all? I'll discuss the rainforest, its value, and what you can do to save it. I'll end with a rundown of great new advances in alternative energies, including solar, wind, and fuel cell technologies.

The appendixes contain further resources and readings for you to explore.

Extras

In addition to the text, you'll find some interesting facts in the following boxed notes:

Climatoids

Here, you'll see startling facts and amazing trivia about our climate.

def•i•ni•tion

Look here for definitions of global warming terms.

Hot Debates

There's lots of debate on this topic—and this is where you'll find it.

Global Warnings

This is where the ominous predictions reside.

Acknowledgments

I want to thank all the people who helped me assemble the facts for this book. They include: Miles Silman, associate professor of biology, Wake Forest University; Robert Jackson, professor of biology and director of the Center on Global Change at Duke University; Dr. Richard Alley, professor of geosciences, Penn State University; Dorthy Peteet, NASA; Kyle Kudworth, Yerkes Observatory; Lynne Talley, Scripps Institution of Oceanography; Dave Whiticre, Peregrine Fund; Brad Boyle, freelance botanist; Louise Emmons, Smithsonian Institution; Tom Schulenberg, Chicago Field Museum; Richard Somerville, professor of meteorology, Scripps Institution of Oceanography; and all the others too numerous to mention that helped me to understand the complexity of this topic.

Also, thanks to my wife, Maggie, for helping me stay sane while pulling this project together. And thanks to my mother, Annabelle, for pointing out those great news stories I missed and for a lifetime of support.

Special Thanks to the Technical Reviewer

The Complete Idiot's Guide to Global Warming, Second Edition, was reviewed by an expert who double-checked the accuracy of what you'll learn here to help us ensure that this book gives you everything you need to know about global warming. Special thanks are extended to David B. Reusch.

Trademarks

All terms mentioned in this book that are known to be or are suspected of being trademarks or service marks have been appropriately capitalized. Alpha Books and Penguin Group (USA) Inc. cannot attest to the accuracy of this information. Use of a term in this book should not be regarded as affecting the validity of any trademark or service mark.

Part 1

Global Warming Signs

So … what is this thing called "global warming"? It's certainly not a love song. Too many people are arguing over it to have anything to do with love. Global warming is either a catastrophe in the making or a great scientific misunderstanding.

In this part, I'll take a look at the basics. I'll run down some of the dire consequences that scientists are predicting if we just ignore the problem. Then I'll take apart the particulars. I'll tell you about the greenhouse effect. I'll get into a little atmospheric chemistry and talk to you about the substances involved.

Then we'll take a look at how scientists have been studying ocean bottoms, tree rings, and polar ice—attempting to get an accurate reading of the problem. I'll round out your education with a four-billion-year history of the atmosphere and help you understand what makes Earth more livable than Mars, Jupiter, or Venus.

"You don't even know where carbon dioxide comes from, do you?"

Chapter 1

A Crisis in the Making

In This Chapter

♦ The inside scoop on rising temperatures

♦ How bad is it going to get?

♦ What's going to change?

♦ Who will be affected?

Not many years ago, scientists were running off to conferences all excited about the possibility that our climate might be changing and the causes could be manmade. There was a flurry of scientific discussion, lots of debates, and numerous dire predictions. The public was mystified: What do they mean? What's all the fuss? Why should we worry? Aren't these natural processes?

Things have changed. Scientists are still running off to conferences, although gloom has replaced excitement. There's less debate and more dire predictions. Global warming has inched up the table of concerns in numerous national and international polls. Yet the topic is still one of the most challenging and complex issues facing the world today. I'll try to make it easier.

In this chapter, I will try to explain some of the different variables in the global warming controversy. Is all the ice melting? What about rising seas?

How will global warming affect our coastline and estuaries? Will hurricanes get stronger? What will global warming do to droughts, fires, floods, and disease? What about wildlife and agriculture? There are a lot of questions.

The View from the Andes

On the crest of the Eastern Andes, about an eight-hour drive on a dirt road from Cuzco, Peru, is an expansive vista of the most diverse forest on earth. On a summer day not long ago, I looked out over the mountains and could see 10 or 20 miles into the Amazon Basin before the view disappeared in the rising mist. Storm clouds boiled up in the pink evening sky, and fog advanced over the foothills. The rain and fog suffuse these mountains with the moisture that makes them so astonishingly full of life.

Miles Silman, a biologist from Wake Forest University in North Carolina, had brought me to this ridge to introduce me to the cloud forests of Peru. They proliferate among clouds born of moisture rising from the Amazon Basin. The constant misting sustains a great variety of trees which in turn support ferns, mosses, bromeliads, and orchids that struggle to lay down roots on any bare patch of bark. It's these plants—plus the wet humus soils, the thick understory, and the immersion in clouds—that distinguish cloud forests from other types of tropical and temperate forests.

Silman and other scientists are attempting to catalog and understand plant and animal life in Andean cloud forests before it's too late. Their task is urgent. Oil companies, having found petroleum and natural gas in the surrounding lands, are cutting roads and laying pipelines through the cloud forest. Native farmers and ranchers harvest the cloud forest for firewood and room to expand.

Biologist Miles Silman explains the threat global warming poses to the Peruvian Cloud Forest.

(Photo courtesy of Michael Tennesen)

Most significantly, the Andes cloud forests are threatened by *global warming*. In other parts of the world, warmer temperatures in the last century have pushed native species toward the poles or altered their growth and migration. But here, the warmth is pushing plant and animal species uphill—and Silman believes they can't move quickly enough. This situation is particularly troublesome since, of 25 biodiversity hotspots (areas with high numbers of plant and animal species) worldwide that conservation groups say deserve special protection, the high Andes is the richest and most diverse by far.

Many of the Andean plants have "shoestring distributions." The area where they can root, grow, and reproduce stretches over hundreds of miles horizontally but minute distances vertically. Says Silman, "I could throw a rock across the elevational range of many different species." These plants' preferred altitudes—and therefore the altitudes of the birds and other animals that feed on them, pollinate their flowers, and disperse their seeds—are determined largely by tem-

def•i•ni•tion

Global warming is the crisis in our climate due to the increase in the average temperature of the world over time created by natural variables and the increase of greenhouse gases, which many scientists believe are caused by man.

perature. And as the Andes cloud forests heat up through global warming, these plants are being evicted from their natural ranges. This occurrence parallels what is happening with the tropical ice caps, which are in full retreat—and most are expected to disappear in the next 50 years.

Who Turned Up the Heat?

Is it really getting warmer? That's what the Intergovernmental Panel on Climate Change (IPCC) says. Established in 1988 by the World Meteorological Organization and the United Nations Environmental Program, the IPCC consists of a group of more than 2,000 scientists. The IPCC concluded in its 2007 report that globally averaged surface temperatures have increased by 1.35°F (.74°C) over the last 100 years. And it also projects that by 2100, our climate will range from 2.0 to 11.5°F (1.1 to 6.4°C) hotter than present. That's a temperature change nearly as dramatic as the one that ended the last ice age.

The principle difference between this report and the one issued by the same group in 2001 is the certainty of its claims. The IPCC now feels that the warming of the climate is unequivocal, and the organization is 90 percent certain that the most of the increase in globally averaged temperatures since the mid-twentieth century is due to the increase in greenhouse gases released by humans.

def•i•ni•tion

The **greenhouse effect** is a natural phenomenon produced by a blanket of gases—carbon dioxide (CO_2), methane (CH_4), nitrous oxide (N_2O), water vapor (H_2O), and others—that act to hold in the reflected heat of the sun which would otherwise escape and warm up the lower atmosphere.

Scientists report that during the last dozen years, we've experienced 11 of the hottest years since 1861—and that's when we started taking the global temperature. However, scientists have used some neat tricks to put together global temperature estimates from tree rings, corals, ice cores, and sediments and believe that the warming of the twentieth century is a record that has not been exceeded in 1,300 years.

What's the cause? Mostly, it's our emission of carbon dioxide (CO_2) into the atmosphere which amplifies the famous *greenhouse effect*, whereby carbon dioxide and a few other gases help retain the sun's heat.

About three fourths of the manmade emissions of carbon dioxide into the atmosphere in the last 20 years are due to the combustion of coal, oil, and natural gas. The rest is predominantly due to tilling the soil and cutting down trees. This removes the vegetation that used to take in the carbon dioxide. Recent carbon dioxide levels, now at about 379 parts per million, have not been exceeded in at least the last 650,000 years.

So what's that mean? Man is becoming a geological force comparable to volcanoes and asteroids—both of which have drastically altered Earth's atmosphere and climate in ages past.

The Ice Is Melting

There have been major changes in global ice fields in the twentieth century. Mountain glaciers have retreated worldwide. Those throughout Glacier National Park in Montana are receding rapidly and within 15 years may be gone.

Since the 1970s, the thickness and area covered by the Arctic sea ice has diminished dramatically. There has been more than a 43 percent decline in late-summer Arctic sea ice thickness, and scientists predict that if things continue as they are, the Arctic ice cap will completely disappear each year during the summer. This in itself will speed warming, because water at the pole will absorb warmth from the sun, not reflect it back as ice does.

The *permafrost* is thawing, which affects travel. Trucks get stuck in the mud when the permafrost melts. Trucks that once traveled on frozen highways in Alaska more than 200 days a year now travel safely fewer than 100 days a year. Trees that once anchored their roots into frozen tundra now lean to both sides, giving them the nickname "drunken trees."

The Greenland ice sheet has experienced a serious decline. Meltwater pools have developed on the surface during summer, and scientists believe that these pools bore holes through the ice that reach all the way to the junction between the ice and the bedrock below. Scientists fear this phenomenon may increase the speed at which the ice flows to the sea.

def•i•ni•tion

The **permafrost** is the vast stretch of permanently frozen subsoil that stretches across the extreme northern latitudes of North America, Europe, and Asia. This land, too cold for the growth of trees, generally marks the northern limits of the forest.

Such meltwater pools and deep, narrow holes in the ice were observed on the Larsen-B ice shelf—an area 150 miles long and 30 miles wide on the Antarctic Peninsula. Beginning on January 31, 2002, scientists watched in shock as the shelf began to break up and within 31 days had disintegrated.

Rising, Warmer Seas

The greatest effects would come with the demise of our great ice masses. If the West Antarctic ice sheet were to go (of which the Larsen-B ice shelf was once a part) or the Greenland ice sheet were to disintegrate, it would raise sea levels worldwide by about 20 feet. This could drown large portions of Florida, including the Everglades. It would devastate the Netherlands and many low-lying island nations. It would displace 60 million people in Calcutta, India, and in Bangladesh. In Manhattan, the site of the World Trade Center would be underwater.

Sea level rise comes from both melting ice and the expansion of the water as it warms. This second effect is important since it will continue as long as the temperatures continue to rise.

Coastlines and Estuaries

Global warming will accelerate coastal erosion, reduce wetlands, and increase the salinity of coastal water sources. Increased salinity would alter the water quality in estuaries, interfering with their roles as nurseries for fish and migratory stopovers for birds. With rising seas, heavier storms, and more floods, populations could be pushed back from the shore.

Rising levels of saltwater could ruin groundwater near tropical and subtropical coastlines as well. Sea-level rise as well as more intense and more frequent storms could thrash coral reefs, salt marshes, and mangrove forests.

Stronger Hurricanes

As ocean temperatures rise, *hurricanes* become stronger. Heat drives hurricanes. When Hurricane Katrina hit Florida on August 23, 2005, it was only a Category 1 storm. But then it passed over the unusually warm waters of the Gulf of Mexico. By the time it hit New Orleans, it was a Category 5, with a destructive force that this city had never seen. Only months before this catastrophe, a major study from the Massachusetts Institute of Technology (MIT) had come to the conclusion that global warming was making hurricanes more forceful and devastating. With sea surface temperatures increasing, we will see more Category 4 and 5 hurricanes in the future.

def•i•ni•tion

Hurricanes are violent storms that originate in the tropics and whose winds exceed 73 mph.

Scientists predict that the intensity of hurricane winds and rain might increase. That means these tremendous tropical storms will get wetter and nastier. Areas like Galveston, Texas, would increasingly be threatened by storm surges—massive amounts of seawater pushed ahead by the intense winds of the hurricane. Storm surges of up to 17 feet created most of the damage and deaths that arose from the Galveston Hurricane of 1900, which took the lives of more than 6,000 people. In terms of the loss of life, it was the greatest natural disaster in American history.

With the intensity of rain inside hurricanes likely to increase with global warming, storms like Hurricane Mitch—which was a rare Category 5 hurricane when it passed over the Gulf of Mexico—may occur more often. Mitch struck Central America in 1998 and killed more than 11,000 people, left 2 million homeless, and caused more than $10 billion in damage across Nicaragua, Honduras, Guatemala, Belize, and southern Mexico. It was one of the worst hurricanes ever to hit the Americas in terms of lives lost. The majority of lives were lost due to massive mudslides created by its torrential rains.

Fire

Hotter summers, particularly over mid-latitude continental interiors, will increase the risk of drought and ratchet up the chances of forest fires. The year 1988—another record year for heat—produced a scorching summer that seemed to set the world on fire. Large tracts of forest in Yellowstone National Park in the United States and the forests in the countryside of France went up in smoke. The fire in Yellowstone burned 1.2 million acres, about 36 percent of the park. Much of Yellowstone is still charred today.

The summer of 2002 was a record fire season in the western United States and Canada as well. A series of fires in Colorado raged through the summer, taking more than 400,000 acres. In California, where climate conditions produce a fire season that generally starts in late summer, the drought got brush, trees, and grass burning in spring. In May, I drove by the Chino Hills State Park in southern California—its grassy mountains all lit up in flames. In July, I backpacked into the Sespe Wilderness—80 miles north of Los Angeles—where fires had closed the roads up until only a few days earlier. On my way home, I watched a fire start on a hillside that erupted with the intensity of gasoline and rushed up the brush-covered side of the mountain. Numerous cars pulled over, and I could see people using their cell phones to call for help.

The searing heat wave of 2006 created a flaming catastrophe throughout much of the western United States. In Montana, the 92,000-acre (38,000 hectares) Bundy Railroad fire forced firefighters to retreat three times in the face of high winds, which fanned burning strands of dried grass, sagebrush, and pines. Fifteen new fires were reported on July 18 alone in Montana, California, Idaho, and other states.

Drought

Water resources are likely to be a hot topic of discussion in arid zones where global warming might lead to a reduction in water supplies. Farmers could see more of their precious irrigation waters routed to growing urban centers rather than to their fields. Cities and farmers along the Euphrates River that flows through Turkey, Syria, and Iraq will experience similar battles as each country promises its people more water than it can deliver. By 2015, according to U.S. and UN estimates, as much as 40 percent of the world's population will live in places where there is simply not enough water to go around.

Floods

Some of the most serious and widespread impacts of global warming will come from landslides and mudslides driven by increased rains and flooding driven by sea level rise. In the mid and high latitudes of the northern hemisphere, there has already been a 2 to 4 percent increase in the frequency of heavy rains over the last 50 years.

Floods in the summer of 2002 raged through Europe. More than 200,000 Czechs were forced to flee their homes during the worst floods in more than a century. The Danube raged too hard for cruise ships, and flooding in former East Germany devastated major towns and tourist centers. German scientists blamed global warming for the deluge. Floods hit Europe again in 2005. Sixty-two people died in Romania, Bulgaria, Switzerland, Austria, and Germany.

In July 2005, 37 inches of rain fell on Mumbai, India, in 24 hours. It was the most rain any city had ever received in a single day in the recorded history of that nation. The water rose to seven feet in some of the flooded areas. More than 1,000 people died from this storm.

Climatoids

Often the point where rainfall becomes a destructive flood is not so much a matter of the amount of rain but how much that rain exceeds the capacity of the land to absorb the moisture. Thus, a season with 30 inches of rainfall in the southwest United States can create havoc—where such rainfall in the Northeast would be tolerated easily due to denser forest cover and the ability of that cover to absorb moisture.

Floods often occur when the ground reaches its saturation point and can no longer contain water. In areas with lots of ground cover, that takes a while—but in deserts where there's little ground cover to contain the moisture, that point is reached quickly. Runoff can turn into a flash flood.

I witnessed the speed and destruction of a flash flood in Baja California, Mexico, when a storm—only a few hours old—sent a torrent of water down a desert canyon. It skipped over the desert like it was on a greased grill with nothing holding it back. The force carved out a portion of Baja's Highway 1, upon which we were headed south. My friend and I turned the car around, but a few miles farther north we had to float it across a river that hadn't been there hours earlier.

Disease

Higher temperatures, more rainfall, and more water pooling on the surface is great if you are an insect—but not if you are a person susceptible to insect-borne diseases. Mosquito-borne malaria could become more common in the United States as mosquitoes move north with warming temperatures. Tick-borne Lyme disease might work its way into Canada. Global warming could also herald the spread of encephalitis and dengue fever.

In general, infectious diseases will show up more often and hang around longer. As with many of the results of global warming, the rise of infectious disease will be felt the hardest by people in Latin America, Africa, and southeast Asia—especially among those countries with the least amount of money to provide care.

In Latin America, insect-borne diseases will expand toward the poles and up the mountains. This has already occurred in Africa. The cities of Nairobi, Kenya, and Harare, Zimbabwe, were formerly located above the altitude of mosquitoes. But as the world has warmed, mosquitoes have moved into these higher places.

Forests and Wildlife

Plants and animals inside our national parks could find that the ideal climate for growth has shifted poleward. The trouble is, plants can't pick up and move to a more hospitable environment—and animals can't follow the habitat poleward—because much of the area surrounding our national parks has been developed by people.

Global warming could have drastic effects on the earth's remaining pockets of wilderness, as well. Forests tend to survive in areas that have the right soil, moisture, and temperature. As the world warms up, forests can move poleward as the wind, birds, and animals transport their seeds. But the pace has its limits. Norwegian scientists studying evidence of the movement of forests northward after the last ice age noted that different forest types sometimes took centuries to establish themselves in areas where the climate had changed to one favoring new forest types.

Increasing carbon dioxide concentrations could actually stimulate the growth of some species. Managed forests—where we get our lumber—will have the easiest time of it because man can help the forest make adjustments.

Dry forests in arid or semi-arid areas, where warming is likely to decrease soil moisture, are expected to diminish. Expected droughts in some parts of the Amazon could boost fires already raging to clear the jungle for crops and cattle. Tropical dry forests in Mexico and northern Africa will become even drier. The same year that Katrina hit the United States, one of the worst droughts in years hit the Amazon. It made scientists aware of how interconnected these climate shifts are.

Global warming pressures will do a number on wildlife as well. Many species already at high risk will be pushed closer to the edge as climate change renders parts of their current habitats unsuitable. Climate pressures in the twenty-first century could cause species currently classified as "critically endangered" to become extinct—and the majority of those labeled "endangered or vulnerable" to become rarer and closer to extinction.

Global warming could alter the habitat for grizzly bears, red squirrels, and other wild-life in Yellowstone National Park by reducing the number of whitebark pine trees—an important food source for these critters. Habitat changes in other areas could pressure other species with local or global extinction. Among the IPCC's list of threatened ani-mals are the forest birds in Tanzania, the resplendent quetzal in Central America, the mountain gorilla in Africa, and the Bengal tiger in India.

Global Warnings _____

The heat has ruined much of the dense, shrimplike krill swarms that Adélie pen-guins feed on in the ocean. Parents now leave hungry chicks alone on the ice while they go out on wide foraging expeditions for diminishing food. Scientists studying the animal warn that the Adélie could be the "canaries in the coal mines" of global warming.

Climate change will lead to loss of habitat for cold- and cool-water fish and a gain in habitat for warm-water species. Inland lakes and rivers are particularly susceptible to climate change. Global warming could eliminate rainbow trout and brook trout in the Northeast and upper Midwest. You might have to go to Canada if you want to use your fly rod (which might not be a bad idea, because red fire ants in the Southeast might expand their ranges into the north).

Polar bears, ringed seals, and migratory birds that depend on the Arctic for a living might feel some of the earliest effects because that area is believed to be one of the fastest warming on earth. Already, Adélie penguins in the Antarctic are suffering some of the most severe effects of present-day global warming.

I'll talk more about the price that wildlife will pay for global warming in Chapter 15.

Agriculture and Industry

In some ways, agriculture might benefit from global warming. Increased carbon diox-ide levels can stimulate plant growth. However, the optimum climate for certain plants might go north as the climate warms from the equator. It's a varied picture: while Canada might have more rain, the Midwest will be 30 to 60 percent drier.

This could mean that ideal weather for some crops, such as wheat, could move north-ward into Canada. But these are rapid evolutions in any plant's cycle—and although the weather may be better in Canada, the soils may not be rich enough to sustain the new crops.

One of the most vulnerable industries is insurance. Insurance companies have seen a 15-fold increase in the amount of money paid to victims of extreme weather in just the last three decades. That's due not only to growing populations but also to increasing development in vulnerable areas and more frequent storms.

Climatoids

In general, global warming will affect rural communities more than urban environments. That's because rural communities often depend upon one or two resource-based industries. If your principal industry is corn, for example, and there isn't enough rain, then it's harder to adapt. People have more options in the city, but city folk won't go unscathed, either. Global warming will probably increase the formation of ground-level ozone and other pollutants.

The Bright Side

There are some positive benefits to global warming. Warmer winters in Canada and the northern parts of the United States may significantly reduce energy bills. Plus, the flowers will blossom sooner and birds will start nesting earlier—even if the summer heat is a killer. Because managed forests will do better, lumber should be cheaper. And tourism may decrease in some parts of the world, but perhaps you don't like tourists anyway. Last but not least, fewer people will be freezing to death.

The Net Effect

Basically, the net effect depends on how warm it gets. A few degrees may not do a lot of damage, but the hotter it gets the more destructive the weather will be. Even for increases of less than a few degrees, the IPCC projects more people will be harmed than benefited by global warming.

The Least You Need to Know

- The average global temperature has risen 1.35°F (.74°C) in the twentieth century.
- Scientists predict the average temperature will rise 2.0° to 11.5°F (1.1° to 6.4°C) in the twenty-first century.
- This warming might lead to increased damage from hurricanes, floods, fires, and disease.
- Global warming will produce major changes for man and nature.
- Although there will be some minor benefits, the net impact will be negative.

2

I Don't See Any Greenhouse

In This Chapter

- ◆ Humans discover the greenhouse
- ◆ The function of the greenhouse
- ◆ Name those gases
- ◆ Why gases are increasing

If all these ominous weather predictions seem hard to fathom, consider that the causes of all this worry are some simple gases that are present in such small volumes that they are measured in parts per million (ppm)—and some in parts per billion (ppb). But although their amounts are tiny, their roles are vital to life on Earth. In this chapter, you'll learn about the discovery of the greenhouse effect, the gases that make up this chemical blanket, and how we humans are disturbing this phenomenon.

Discovering the Greenhouse Effect

French mathematician Jean Bapiste Joseph Fourier first noted the greenhouse effect in 1827 when he compared the atmosphere to a "glass vessel." Fourier was the first to recognize that some of the gases in our atmosphere act like a greenhouse because they let in sunlight—like a greenhouse does—but prevent some of the sun's warmth from radiating back out into space.

In the 1850s, British physicist James Tyndall tried to figure out which of these gases contributed the most to the warming effect. He measured the heat-trapping ability of various atmospheric gases and was surprised at the results. It seems that the two most common gases, nitrogen and oxygen, which make up 99 percent of the dry atmosphere, have no heat-trapping ability at all. Tyndall concluded that the job of keeping the world warm and toasty fell to carbon dioxide, methane, water vapor, and a few other elements.

Climatoids

The greenhouse gases—carbon dioxide, methane, and others—are long-lived gases (carbon dioxide lasts 100 to 200 years in the atmosphere; methane 9 to 15 years) that are found in minute quantities within the atmosphere. Water vapor is short-lived and alters their effects. Although the long-lived greenhouse gases play a vital role in keeping the planet warm, they make up less than 1 percent of the total atmosphere. No wonder so many people have such a hard time fathoming the dangers of their imbalance. They are the true phantoms of the atmosphere.

In the 1930s during a period of particularly hot summers in Europe, George Callendar, a British coal engineer, put together several decades worth of temperature readings taken at weather stations and from shipboard decks and uncovered that the world was gradually getting warmer.

Callendar published his results in 1938. He attributed this warming to the rising levels of carbon dioxide brought on by the burning of fossil fuels. He was way ahead of his time. However, just after he came out with his results, Europe went into a prolonged cold spell—and nobody seemed to care.

Besides, scientists felt that the oceans, which acted as vast carbon sinks, would absorb excess carbon dioxide. Excess carbon dioxide would get sucked in by the seas—and by the plants and animals therein—and end up deposited on the sea floor. It wasn't until the 1950s that scientists at the Scripps Institute of Oceanography found that seawater had a limit to its absorption. They projected that, as increasing amounts of carbon dioxide were added to the atmosphere, the ocean would reach its saturation point and would no longer act as a ready sponge for humanity's excesses. Although much of the gas would end up in the sea, at the rate humans were adding it to the atmosphere, there would be more than enough left over to influence the climate.

Observations since the mid-1900s show that the ocean is absorbing 80 percent of the global warming currently experienced and that the average temperature of the Earth's oceans has increased to depths of 3,000 meters (9,842 feet). These findings occurred about the time that plans were being made for the International Geophysical Year (July 1957 to December 1958), and the carbon dioxide issue ended up on its agenda. The event was a vast and ambitious project that lasted more than 18 months and employed the help of geologists, chemists, climatologists, and other specialists from more than 79 nations. It was to be Earth's first physical checkup, and scientists were sent all over the world to measure everything they could.

The Keeling Curve

One young scientist sent out on this monumental task was Charles David Keeling. He was fresh off research that had him running around California stuffing samples of the air into bottles that he took back to the lab to measure. Keeling developed a device for measuring carbon dioxide in the atmosphere—the first ever to be able to measure one part per million. He noticed that his samples were 315 ppm of carbon dioxide, which he would later learn was about 13 percent higher than before the Industrial Revolution.

The Industrial Revolution is a term applied to the era between 1750 and 1850 in Britain, when English society transformed from a stable agricultural economy to a modern industrial one. The biggest changes in the United States followed the Civil War from 1860 to 1890. Other countries have had their industrial revolutions in different periods, and some third-world nations are still waiting for the change. From a greenhouse gas perspective, the world's carbon dioxide concentrations started their greatest rise after 1850.

Keeling decided to expand his research. For the International Geophysical Year, he built two instruments that he called manometers, which were able to take continual readings of the carbon dioxide in the atmosphere. He flew one to Mauna Loa, a high mountain volcano that towers over the big island of Hawaii. He felt that out in the middle of the ocean, high up on a volcano, far from any city and high above the Hawaiian rainforest, he could get what he thought was a pure, undistorted reading of the earth's carbon dioxide content.

He began his readings in 1958. At first, the result was a marvelous record of plant life on our planet. The figure on his chart was a series of humps that represented the seasons. Carbon dioxide was high in the winter, falling in the spring, low in the summer, and rising again in the fall. What it was recording was the annual, primarily Northern Hemisphere global plant cycle. In the increasing warmth of the spring and summer, plants burst into growth—producing leaves, flowers, and fruit—and in that growth process, *photosynthesis* was removing carbon dioxide from the atmosphere. Then in the fall, plant leaves, flowers, and fruit dropped to the forest floor. As they disintegrated, the carbon dioxide was released back into the atmosphere.

def•i•ni•tion

Photosynthesis is the process in which green plants take in carbon dioxide (CO_2; mostly through their leaves) and water (mostly through their roots) and mix them together in the presence of sunlight and chlorophyll to produce glucose, a carbohydrate utilized by the plant as energy.

Keeling was fascinated with these cycles, so he kept his machine running long after the initial research project. Over several decades, his original manometer and a number of others like it were set at various stations around the world to keep track of the atmosphere's volume of carbon dioxide—much like you might keep track of your cholesterol. But when Keeling started collating the data and plotting it onto a chart, he uncovered a startling fact. The result is what we now call the Keeling Curve. Although each year had its cyclical bumps, the overall trend was an unmistakable rise upward.

Prior to the Industrial Revolution, there were approximately 280 ppm of carbon dioxide. In 1958, there were 315 ppm. In 1990, there were 355 ppm, and in 2005 there were 379 ppm. That means there is 35 percent more carbon dioxide in the atmosphere now than there was before the Industrial Revolution—and a 20 percent jump in just the last 50 years. In the 1960s, Keeling found that the carbon dioxide levels in the atmosphere were rising at 0.7 ppm per year. In the 1980s, the annual rise was up to 1.5 ppm—more than double. From 1995 to 2005, the annual growth was 1.9 ppm per year.

Hot Debates

We spew about 4 tons of carbon dioxide into the atmosphere for every one of the 6 billion men, women, and children on Earth. Americans contribute closer to 20 tons each. They consume about 22 percent of the world's oil although they make up only 5 percent of the world's population.

The trouble was, most of Keeling's work was done prior to 1970 when much of the world was in a cold snap. Although carbon dioxide levels steadily rose over the course of the century, temperatures in general rose until the 1940s but then dropped back down until the 1970s, when they began rising again toward new records. Ironically, it might have been our own pollution that kept things cool for those middle decades. The National Oceanic and Atmospheric Administration (NOAA) estimates that the surge in coal and oil use during and after World War II sent increasing amounts of sulfate particulates into the air, which cooled the atmosphere down. However, around 1970 the increase in the longer-lived carbon dioxide and methane overwhelmed the shorter-lived sulfates—and the temperature started to rise again.

Its Vital Role

The greenhouse effect is not a bad thing. In fact, we need it to stay alive. It helps regulate the temperature of our planet. There are a number of chemical compounds in Earth's atmosphere that make up the greenhouse gases. Greenhouse gases effectively trap enough of the sun's reflected heat to raise the earth's temperature to what it is today. Without a natural greenhouse effect, the average global temperature of our planet would be around –2°F (–19°C) instead of its present 58°F (14.5°C).

How would you like to roast your hot dogs in –2°F weather? It wouldn't be too comfortable. In fact, life as we know it wouldn't exist. So we need the greenhouse effect. No one is arguing that it isn't important. The only disagreement revolves around whether humans are increasing the atmospheric content of these gases and what consequences that will bring.

Greenhouse gases allow direct sunlight, which is relatively short-wave energy, to pass by them unaffected. That's because greenhouse gases are poor absorbers of short-wave energy. It's just not their cup of tea. But as the short-wave energy heats up Earth's surface, it radiates back in the form of longer-wave energy—which is like pizza and ice cream to greenhouse gases. They gobble up that heat (which would otherwise be lost to space) and reflect some of it back downward again. This warms up the lower 10 miles of the atmosphere (where we live, work, and play basketball).

Many of these greenhouse gases can occur naturally in the atmosphere, such as carbon dioxide, methane, water vapor, and nitrous oxides. Others, such as chlorofluorocarbons (CFCs), are manmade. These are the gases that once cooled our refrigerators.

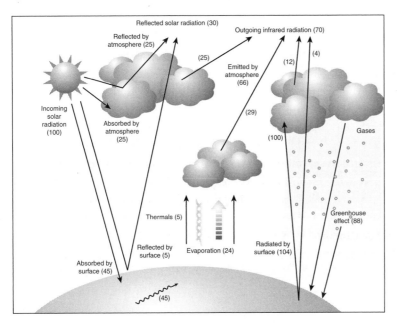

The greenhouse effect is very real.

Carbon Dioxide (CO_2)

There are a number of greenhouse gases, but carbon dioxide seems to gather the most attention because it is long-lived and causes the greatest amount of global warming of the long-lived gases. We know from chemistry classes that carbon dioxide is one of the basic substances of life. We breathe it out, and plants take it in. It's the bubbles in your cola, your mineral water, and your beer. Pretty innocuous, wouldn't you think? How could it be such a problem?

Where It Comes From

Most mammals—including dogs, cats, birds, deer, mountain lions, you, and I—create carbon dioxide when we exhale. Plants take it up and use it in photosynthesis to make leaves, flowers, and fruit. But plants also give off carbon dioxide in fall when those leaves gather on the forest floor. As they disintegrate, the carbon dioxide they took up with photosynthesis is released back into the environment.

Volcanoes also give off carbon dioxide. In fact, this is how the planet got its first atmosphere. Our first air on Earth was a blanket of carbon dioxide from ancient volcanoes. It was only through the evolution of plant life, which began to take up the carbon dioxide and exhale oxygen, that the planet's atmosphere changed into one that was mostly oxygen. This condition then allowed for the evolution of animal life. (For more details, see Chapter 3.)

The concentration of carbon dioxide in the atmosphere has increased about 31 percent since 1750 and 18 percent in just the last 40 years. IPCC scientists report that the present amount of carbon dioxide has not been exceeded in the last 650,000 years.

Where Is It Stored?

I talked about how plants store carbon dioxide. The atmospheric concentration of carbon dioxide has a marked seasonal shift in the northern hemisphere. That's because the northern hemisphere has more land and more vegetation covering it. Remember Keeling's manometer and how the graph went up and down with the changing seasons? That seasonal roller coaster is not as pronounced in the southern hemisphere, which has more water. Carbon dioxide is also stored in the oceans as well as in rocks.

What's the Problem?

The problem is that carbon dioxide is on the rise in the atmosphere, and that's having an effect on the global average temperature. Prior to the Industrial Revolution, concentrations of carbon dioxide in the atmosphere were relatively stable at about 280 ppm. Today, the concentration is about 379 ppm and rising—an increase of about 35 percent. Concentrations have risen about 18 percent in just the last 40 years. What makes the problem so unique is that carbon dioxide has a life span in the atmosphere of about 100 to 200 years, which means that even if we stopped burning fossil fuels right now, the effects could last for centuries. The IPCC, in its 2007 report, warns that warming and sea level rise could continue for centuries due to climate processes and feedbacks—even if greenhouse gases are stabilized.

By the year 2100, IPCC scientists predict that the atmospheric concentrations of carbon dioxide could rise from current levels to a range of from 490 to 1,260 ppm (75 to 350 percent above preindustrial figures) if we don't stop pouring this gas into the atmosphere. About three fourths of these emissions are due to the burning of oil, coal, and natural gas. Some of that gas comes from the more than 500 million cars on the planet. Some of it comes from the fires used to clear the land and create farms and cattle ranches.

Some of it comes from cooking fires. It is estimated that more than two billion people use firewood to fix their daily meals. Open fires are not only inefficient sources of energy but significant contributors to atmospheric carbon dioxide.

Climatoids

Since 1900, the average annual consumption of energy has increased more than tenfold. Between 1870 and 1970, the burning of fossil fuels to create this energy added 400 billion tons of carbon dioxide to the atmosphere. From 1970 to 1989, another 400 billion tons were added. The increase is due to the tremendous acceleration in our consumption of coal, oil, and natural gas in the last half of the twentieth century.

We can't just plant trees to get rid of all the extra carbon dioxide. Even if we could put back all the trees we cut down, carbon dioxide in the atmosphere would only be reduced by 40 to 70 ppm. Eventually, we will have to reduce the amount of carbon dioxide emissions if we are to bring it under control. Carbon dioxide emissions will have to decline to a small fraction of what they are today in order to stabilize the atmosphere over the long run.

Methane (CH₄)

You might think of methane as the wool blanket of greenhouse gases. That's because it is an extremely efficient absorber of heat. It's 20 to 30 times more efficient at trapping heat than carbon dioxide. Still, methane concentrations in the atmosphere are below 2 ppm—and its lifetime is only about 9 to 15 years. However, that concentration is still about 150 percent higher than it was before the Industrial Revolution. The amount of methane in the atmosphere in 2005 (1,774 ppb) exceeds by far the natural range of the last 650,000 years.

Climatoids

As human populations continue to grow, so do their cattle. There are over a billion head of cattle on the earth today. Cows produce methane when bacteria break down cellulose in their stomachs. The gas comes out either end of the cow.

Methane, also called swamp gas, is the principal component of natural gas. It has both natural and human-influenced sources. Naturally, it's given off as part of the decomposition that occurs in wetlands, swamps, and bogs. Humans have added a host of other sources to these emissions. They include the bacterial decomposition that occurs in landfills, rice paddies, and the guts of cattle and termites. As humans cut down the forest, termites and bacteria flourish in the refuse left on the forest floor.

Some methane also escapes during the process of coal mining and oil drilling. Methane is also emitted during the incomplete combustion of oil, gas, and coal. Methane is stored in the vast soils of the Arctic tundra, which is one of the reasons (along with increased CO_2 release) that scientists are concerned about the warming of the permafrost. It could release enough methane to drastically affect the climate. There are also huge reserves of methane in the sediments at the bottom of the ocean floor. Because of its efficiency as a greenhouse gas, it wouldn't take much additional release to have an effect on the atmosphere. Methane increases in the atmosphere have slowed in the last 10 years, although scientists are at odds to explain why.

Nitrous Oxide (N₂O)

Like the other greenhouse gases, nitrous oxide has been on the rise since the beginning of the Industrial Revolution. Because nitrogen makes up 78 percent of the atmosphere and oxygen another 21 percent, it's easy to get these two chemicals together. Wherever fuel is burned, it can force nitrogen and oxygen to combine to make nitrous oxide. It's part of those pesky gases that make up photochemical smog, occasionally found in the air over Los Angeles and other major industrial cities.

Microbial processes in soil and water also get these guys going, especially if you toss in some nitrate fertilizers. Nitrate fertilizers have been increasingly used over the last century.

Ozone

There is a broad band of ozone in the stratosphere. Only small amounts fall out into the lower atmosphere, but in the twentieth century we've added to that lower-atmospheric concentration. Ozone in the lower atmosphere has increased by about 30 percent since the preindustrial era.

Chlorofluorocarbons (CFCs)

These gases have no natural sources. Chlorofluorocarbons were synthesized in a laboratory in the late 1920s and were initially heralded for their many uses, particularly as refrigerants. But they were also used as aerosol propellants and cleaning solvents. However, scientists discovered that they gradually accumulated in the environment and were able to destroy ozone in the upper layers of the atmosphere. Thanks to an international agreement, the levels of the major CFCs have peaked and are now declining.

Carbon Monoxide (CO) and Others

The carbon monoxide concentration in the atmosphere influences the greenhouse effect because it can modulate the production of methane and ground-level ozone. By itself, carbon monoxide does not absorb heat from the sun as do the other greenhouse gases. The northern hemisphere contains about two times the amount of carbon monoxide as the southern hemisphere, partly because people in the north drive a lot more cars.

Carbon monoxide is one of the gases that may be on the wane. Measured concentrations in the air were generally increasing until the late 1980s, when they began to decline. Many scientists believe this is due to the increased usage of catalytic converters on automobiles.

It's colorless, odorless, and tasteless, but it is harmful to people and animals because carbon monoxide substitutes for oxygen in the blood. The blood is actually more attracted to carbon monoxide than oxygen (although our bodies need oxygen, not carbon monoxide). Carbon monoxide poisoning can begin at concentrations of 20 to 50 ppm. Cigarette smoke contains 400 to 450 ppm. In a busy intersection, concentrations are about 100 ppm. That might be why you have a headache.

Volatile organic compounds (VOCs) are also greenhouse gases, although their effects are small compared to the others. Their largest source of emissions is from natural vegetation, but they are also a byproduct of automobile exhaust and fuel production. Measuring them is extremely difficult. Still, they are thought to be on the increase in the last part of the twentieth century.

Humanity's contribution to the atmosphere.

(Photo courtesy of IPCC)

Indicators of the human influence on the atmosphere during the Industrial Era

(a) Global atmospheric concentrations of three well mixed greenhouse gases

(b) Sulphate aerosols deposited in Greenland ice

Particulates

Particulates are very small airborne particles or droplets that can influence the climate in a variety of ways. Particulates can absorb the heat radiating off the earth's surface much like greenhouse gases. They also have an effect in clouds, where moisture coalesces on them to create rain. Thus, the life of a particulate in a cloud is rather short—typically no more than a week—which is why particulates are found close to their sources. Unlike carbon dioxide and other greenhouse gases, they don't distribute evenly over the globe.

Particulates are produced in a variety of ways—including the dust from a dust storm, salt from saltwater spray, and the ash from volcanic eruptions. They're also in that soot coming from coal-fired smokestacks. And they're in the ash from forest fires. Most particulates are found in the lower atmosphere, but some occur higher up as well. Volcanic eruptions can send particulates in the form of volcanic ash into the upper atmosphere. Several explosive eruptions that have occurred over the last 150 years have had temporary cooling effects on the atmosphere. The particulates from an increase in coal-fired plants following World War II may have caused a mid-century cooling.

Global Warnings

Particulates can make it into the upper atmosphere, either through a volcanic eruption or the rarer impact of an asteroid with the earth. Despite the Hollywoodization of this threat, many scientists still think that the impact of an asteroid with the planet some 65 million years ago might have created such a blanket of dust (particulates) that it blocked out the sun. And this may be why the dinosaurs went extinct.

Water Vapor

Although we saved water for last, it's not because it's the least. Water vapor is actually the most abundant greenhouse gas in the atmosphere. But its effects are more transitory. Perhaps it's better to think of water vapor more as a modulator of the effects of the other longer-lived gases. Water vapor is a principle driving force for weather. It's also an important feedback mechanism, because as the temperature of the atmosphere rises, more water evaporates from lakes, rivers, the soil, and the sea. The hotter it gets, the more water the atmosphere can hold.

Water vapor is the most abundant of the greenhouse gases, but scientists still debate its contribution to global warming. As a greenhouse gas, water is an efficient absorber of radiated heat (which can warm the planet significantly). But more water means more clouds, and clouds can act either as a blanket over the land or as a reflector of solar energy. Which will be its dominant influence in the future is uncertain.

The Least You Need to Know

- The greenhouse effect is real, and we need it—otherwise, we'd freeze.

- Carbon dioxide, methane, nitrous oxide, chlorofluorocarbons, and some other long-lived gases may be adding to the effect.

- Water vapor acts as a greenhouse gas, but it's short-lived.

- Human activities are upsetting the delicate balance of these elements in the atmosphere.

Taking a Reading

In This Chapter

- How we record the world's temperatures

- Extracting ancient air from Greenland ice

- What else the ice tells us

- Sea bottoms, tree rings, and other ways to read the past

To figure out how to tell the future, you need to be able to read the past. The past tells how temperatures have changed and what the causes and effects were. Scientists who are trying to predict the future run their computer scenarios on the past to see whether it gives them the recorded results. This is the way they calibrate their prediction models. The trouble is, we haven't been taking the temperature that long. So scientists have had to come up with a few sly methods to tell us what the weather was like back then.

The Historical Record

Did you know that moms weren't able to read their kids' temperatures before Galileo? The same guy who popularized the telescope invented the thermometer in the early 1600s. The barometer and rain gauge followed a little later. The big problem was standardization. One scientist, comparing and calibrating various types of thermometers in the late 1700s, listed no fewer than 77 different scales in use then.

Historical temperature records stretch back to about 300 years in a few places, but for much of the world it has only been a few decades to maybe a century. The longest series of historical measurements ever assembled was by Professor Gordon Manley, a British scientist. That record gives the mean temperatures for every month from 1659 to the 1970s at a typical lowland site in central England.

Earth's temperature for 140 years.

(Photo courtesy of IPCC)

Trying to guess the weather without a thermometer takes a little more detective work. Grain prices tell you about the weather, because the price of grain corresponds to the yield of the latest harvest. Diaries and chronicles help at times but often don't add much detail. A fourteenth-century scroll from the manor of Knightsbridge, London, refers to a great drought in the summer of 1342. Medieval monastic records make various references to great floods, droughts, and shipwrecks. Often, the weather is only mentioned in township chronicles as an excuse for spending more on bridge or harbor repairs.

The oldest year-by-year record that comes down to us is the flood levels of the Nile River in Lower Egypt, which gives us an idea of the summer monsoon rains back then. Some records inscribed in stone go back to the first dynasty of the pharaohs (about 3100 B.C.E.).

Some of the best-kept records of temperatures have come from ships. Sailors have a personal interest in trying to figure out the weather. Early sailing vessels would take the temperature by putting a bucket over the side, pulling up some water, and tossing in a thermometer. As with many early readings, scientists today have to allow for errors. While you're measuring the temperature in a bucket, there is water evaporating from the side of the bucket—which has a small effect on the temperature of the water in the bucket. Scientists studying ship records have had to go back and find samples of these early buckets, study them, and make adjustments to their readings to get accurate measurements.

It's the same on land. If you put your thermometer in the middle of a town and a city grows up around it, that affects the temperature. It's known as the *urban heat island effect*. Scientists have worked hard to take those effects out of their calculations about the weather. You can't just take a temperature measurement from old sources and one from a new source and compare the results. You must go back and work on the variables that affect accuracy.

def•i•ni•tion

The **urban heat island effect** is the effect whereby temperatures taken in urban areas, or cities, are higher than in surrounding rural areas. This is due to the heat-absorptive qualities of cement, asphalt, and other building materials, coupled with all the heaters running.

Taking the Earth's Temperature

Satellites help determine the temperature of the earth today. They measure the amount of radiation that the earth is sending back to space. If you have an electric stove and turn it on low, you don't see it too well. But if you turn it on high, it glows bright red. The planet works the same way. The satellite can look down and determine how bright the glow is. The warmer we are, the brighter we glow.

The thermometric record of the earth's temperature has come under fire because skeptics say that scientists have not properly taken into consideration the urban heat island effect. The 2007 IPCC report, however, indicates that although urban heat island effects are real, they are local and have a negligible influence over land and zero over the oceans.

One thing they look at is *glaciers*. Glaciers are weird beasts. They surge, and their ends wiggle—but if you watch a whole lot of glaciers for a whole lot of years, they respond to temperature more than anything else. And if you look at records of glaciers over the last 100 years, almost every glacier on earth has gotten smaller.

def•i•ni•tion

Glaciers are moving masses of ice and snow that accumulate in areas where the rate of snowfall exceeds the rate at which the snow melts.

There is another effect that global warming scientists use to prove their case. If you take a roast out of the freezer and put it into the oven, then the outside will warm up while the inside will lag behind. If you were to drill a hole and put a thermometer in the roast, you'd find the outside is warm and the inside is cold. That tells you the roast is changing from cold to warm. If the roast had been in the oven long enough, the outside and inside would be the same temperature.

It's the same with Earth. If you drill a hole in the earth and you drop a thermometer 100 meters down, the average yearly temperature is a little bit colder than the surface in most places on earth. This is telling us that our planet is warming up. The core of the earth is really hot. It's not a natural phenomenon for the earth to be colder below the surface. It means that the surface is driving the temperature differential, that surface temperatures were colder in the past and are now warming up.

And the numbers you get from the tree rings, ice cores, thermometers, glaciers, satellites, and from below-ground temperatures all say the same thing: the earth is getting warmer.

The Ancient Record

To read the history of the earth's temperatures, scientists look at ocean bottoms, lake bottoms, ice sheets, coral reefs, and tree rings. The layers in these objects represent a history book of the past. Sediments in ocean bottoms, lake bottoms, and ice sheets come from snow, rock, plants, pollen, and the skeletons of living organisms that have accumulated—layer upon layer—over the ages. Corals and tree rings give us a history of the weather and the health of the ecosystem. Scientists looking at these things are doing the same thing as archeologists when they peer through trash dumps of ancient cultures. They are looking for clues to the past.

Ice Cores

One focus of the International Geophysical Year of 1957–1958 was the little-explored icy regions of Greenland and Antarctica. A number of coring projects started where scientists drilled into the ice and extracted long, frozen cores. But the initial cores were taken in less-than-ideal places, so in 1989 an international consortium was established where representatives from Europe and the United States drilled holes 20 miles apart at the summit of the Greenland ice sheet—the enormous glacier that covers most of Greenland. Although a portion of the world's permanent ice is locked into glaciers in high mountain regions around the world, fully 99 percent of the earth's ice is in the great ice sheets that cover most of the island of Greenland and nearly the entire continent of Antarctica.

Richard Alley, Evan Pugh Professor of Geosciences, Penn State University, was part of the Greenland Ice Sheet Project 2—the U.S. contingent of the ice core research. He describes Greenland as a "wonderful place but very cold. The average temperature on the dome was –31°C." In Greenland, the temperature gets up above freezing only about one day per century—and scientists can actually see this in the core samples they take from the ice as areas of snow melt.

The midnight sun circled the horizon as the scientists worked to bring up their samples. Most of the workers wore thick, insulated jumpsuits like you might see workers wearing in an ice cream factory and lived in tents or canvas Quonset huts when not working. The scientists worked alternate 6-week shifts through the summer, although the drillers hung around for 12 to 13 weeks.

The ice sheet that lies over central Greenland is an accumulation of compacted snow turned to ice that is about two miles thick. About 10 percent of the volume of glacial ice is air—which, when compacted, gathers into little bubbles that are samples of the atmosphere at approximately the time the ice was formed. Alley studies ice cores from the past to give us clues as to how the temperature is going to change in the future.

Global Warnings

If all the ice on Earth were to melt right now, the global sea level would rise 200 feet. That would be enough to move the coast of Florida somewhere up into Georgia.

To get at those samples of air, he and his associates drilled down into the ice with equipment that looked very much like an oil derrick, including a 100-foot tower, miles of pipe, and a large geodesic dome around the base that kept the weather off the platform.

Teeth were cut on the end of the eight-inch-thick pipe so that as the drill descended, it would capture a sample core of ice in the middle of the drill pipe. The core samples that came out of the drill were about 20 feet (6 meters) long. Alley and the other scientists took those down to samples of $6\frac{1}{2}$ feet (2 meters) for analysis and then 3 feet (1 meter) for shipping. Samples were sent via refrigerated airmail to the National Ice Core Laboratory in Denver, a project supported by the National Science Foundation and the United States Geological Survey.

So ... What Have We Got?

A lot of the analysis, however, was done in Greenland. Scientists took an immense snow blower and cut a 20-foot-deep trench in the snow, put a lid on it, and created a subsurface lab right on the summit of the ice sheet. As the cores were brought into the lab, the scientists made lengthwise slices of the segments to be analyzed by a variety of chemical, electrical, and visual methods.

When the cuts were made, the different scientific groups split up—taking their samples for further analysis into little rooms cut off the trench. One of the processes used was to take two electrodes and draw them down both sides of the length of the core. This measured how electricity flowed from one electrode to another through the ice. If there was a lot of volcanic acid in the ice, electricity flowed well. If there was a lot of dust in the ice, electricity didn't flow well at all.

In central Greenland, the snowfall accumulates on the summit. And because the temperature only very rarely rises above freezing, the snow doesn't melt but instead accumulates and is compacted into ice from the weight of the snowfall above. About 12 inches (30 centimeters) of ice represents a year in the top portions of the ice sheet. But as the layer gets buried deeper and deeper in the ice, it gets more compacted. When the layer gets buried halfway through the ice sheet, the layer gets stretched and thinned to 6 inches (15 centimeters) thick. When the layer is buried three quarters of the way through the ice sheet, the layer is only 3 inches (7.5 centimeters) thick. Compaction increases as it goes deeper to the point where it becomes difficult to read annually.

Climatoids

Greenland contains air bubbles that are samples of the ancient atmosphere. The bubbles are compacted as they sink deeper into the ice, and according to old polar hands, the ice is great for cocktails because of the loud pops the ice makes when you pour alcohol on it.

Although scientists use various chemical and electrical methods to determine the age of the section, Prof. Alley and his associates were also able to determine annual layers in the ice by putting them on a light table. The low-density, coarse-grained layers formed by the summer sun show up as light bands while the fine-grained snow packed by winter storms and winds appears darker. When they checked their estimates of age against the fallout of historically dated volcanoes, which appear in the ice as layers of ash, they found they had it right.

Inside the ice cores are samples of the air taken from near the time when the snow first fell. The air in the ice collects into bubbles and is further compacted, but upon release it is a fairly accurate representation of the atmosphere at approximately the time the ice was formed.

Scientists can't take measurements of the rainfall and temperature in core samples, so they use *proxies* to help them out. Determining the varying amounts of oxygen *isotopes* is one proxy method that tells scientists the temperature of the atmosphere when the rain fell.

One of the isotopes that scientists look for in Greenland ice is oxygen-18. Oxygen has 8 protons and 8, 9, or 10 neutrons that combine to make oxygen-16, oxygen-17, or oxygen-18. When hydrogen combines with oxygen-18 to make H_2O (water), it forms a heavier water than hydrogen combined with oxygen-16 or oxygen-17 would make. Although this heavier water constitutes less than 1 percent of the water, all of these water samples are a little heavier—and the difference can be measured.

def•i•ni•tion

Proxies are methods of determining values such as temperatures and rainfall by using substitutes, which give indirect measurements. **Isotopes** are varying forms of an element that have closely related properties but different atomic weights.

When a raindrop forms in a cloud, the cloud feeds water to the droplet. And because oxygen-18 is heavier, it falls out of the clouds faster. Scientists find a higher ratio of oxygen-18 in the raindrops and a lower ratio in the clouds.

Because cold air holds a lot less water vapor than warm air, clouds moving from the tropics toward the poles lose water constantly. And the farther north the cloud goes, the more it gets depleted of oxygen-18. It turns out that if you measure the amount of oxygen-18 in snow or water, it is indirectly proportional to the temperature when it fell. Scientists can use the measure of the oxygen-18 content to determine the temperature when the snow fell in an ice core, when a tree took up water, or when shellfish incorporated the water into their shells. This is important, since scientists use these measurements to get an accurate picture of our climate history and that relates to global warming.

A Varied Record

Greenland is an excellent place to study ice because it collects evidence of the earth's atmosphere from halfway around the world. It gathers snowfall from Greenland and collects dust blown to it all the way from Asia, sea salt from the ocean, and methane that comes to it from swamps in the tropics.

Climatoids

The winds of the world do a good job of mixing the atmosphere. If you were to release a large quantity of gas in Seattle, it would only take a few years before people in Bangkok, Thailand, could measure what you released. Dust and sea salt don't mix as well, but most gases are in the atmosphere long enough to mix globally.

Methane is among the many gases that interest Greenland ice core scientists. Methane prior to man taking over the earth was essentially formed from swamp gas. Huge changes in swamp gas tend to go right along with changes in dust and temperature. If there is a lot of methane in bubbles from gas core samples, it means that at the time those bubbles were made, there were a lot of swamps in the world. Determining whether those swamps were in the tropics or in Siberia is another part of the riddle.

To solve this riddle, scientists also measure the methane in air trapped in ice cores in Antarctica. Because methane has a limited life span in the atmosphere, the farther it gets from the source the less methane will be in the atmosphere. So to determine where the methane in the ice is from, you have to compare it with the methane in the ice from Antarctica. If the methane is from the tropics near the equator, then there will be an equal amount in both Greenland and Antarctic ice. If the methane is from Siberia, there will be more methane in the Greenland ice—because Greenland is closer to Siberia—and less methane in the Antarctic ice. Most methane in the ice is from the tropics, and its presence tells scientists how extensive the tropics were during the period that the bubbles represent.

Carbon Dioxide in the Ice

By having samples of the ancient atmosphere, scientists in Greenland have been able to measure the amounts of carbon dioxide in the atmosphere in historical times. Because carbon dioxide is the main greenhouse gas responsible for global warming, scientists have looked in the ice cores to investigate its history.

What they've found is that over the last 110,000 years, the concentration of carbon dioxide has never been nearly as high as it is today. In 1750, at the beginning of the Industrial Revolution, the concentration was about 280 ppm. The present concentration is about 379 ppm and rising—an increase of about 31 percent and a 20 percent jump in just the last 50 years. That rise is unprecedented in the last 10,000 years. If you go back before 1850, the concentration of carbon dioxide in the atmosphere has remained steady over time. It's always been about 280 ppm. This fact adds credence to the assumption that the current rise in carbon dioxide is derived from human causes, because 1750 is around the time that people started burning fossil fuels.

If you go back even further—before 10,000 years ago—there are some variations. During the last ice age, carbon dioxide concentrations were about 200 ppm. It varied with the ice ages as they came and went approximately every 100,000 years. During the last four glacial cycles over the last 650,000 years, carbon dioxide has gone from 280 to 200 ppm when it got cold and then back to 280 ppm during the interglacial warming periods. It never got above 280 ppm.

Hot Debates

Scientists believe that the decrease in the level of carbon dioxide from 280 to 200 ppm probably didn't cause the ice ages. The decrease was caused by the earth's orbit around the sun. But the orbit can't explain all of the climate change. The carbon dioxide change amplified those changes because as a greenhouse it didn't do its job of holding in the sun's radiation. During the ice age, this made the cold parts much colder.

Why the carbon dioxide dropped during the ice ages is a mystery. Some scientists think it has something to do with a reorganization of how the oceans circulate, particularly around Antarctica where the deep ocean comes up to contact the surface and surface waters descend to take their place. It's thought that if you change this circulation, then one of the results might be that the plants in the ocean would do a more efficient job of removing carbon dioxide. If you change the way the waters circulate, you may increase the amount of carbon dioxide taken from surface waters.

Analyzing the Climate Clues

Greenland ice is not the only method scientists use to study ancient climates. There are a number of others—including tree rings, lake bottoms, swamp bottoms, sea bottoms, and coral samples—that give scientists a look into the past. Scientists use measurements from these sources to provide additional information and also to double-check original sources.

Climatoids

At the Laboratory of Tree Ring Research of the University of Arizona in Tucson, scientists have built a chronology of ring widths from bristlecone pine trees in the White Mountains on the California-Nevada border. Those trees are some of the oldest trees on Earth, and University of Arizona scientists have constructed a tree ring record that goes back almost 9,000 years to 7000 B.C.E.

Tree Rings

Have you ever counted the annual rings of a tree stump to figure out how old the tree was? The thickness of the tree rings tells scientists something about the climate and how happy the tree is with climate change. If it's in a place that's cold, it will tell you it's happy when it's a little warmer. And if a tree is in a place that's really dry, it's happy when it's a little wetter. ("Happy" equals a thicker tree ring.)

Fortunately, we don't have to cut the tree down to get at the climate record in its growth rings. Scientists can extract this record by taking a pencil-thick core from the tree without stunting the tree's growth. Ring widths in trees that occur at the upper altitudes or poleward limit of the tree's range give scientists the best picture of the climate. If the ring is thick, it probably indicates a warm year. If it's thin, it indicates a cool summer. In trees that are found in the warm, arid margins of their natural extension, a thick ring could indicate good rain whereas a narrow ring could indicate drought.

Lakes and Peat Bogs

Sediments in lakes and peat bogs give scientists their best clues as to the plant vegetation in a given area. The layers cough up pollen samples or beautiful fossilized leaves of ancient plants. These give scientists clues to the atmosphere because they know what climate the plants prefer.

One of the first histories of the postglacial renewal of vegetation was written in 1876 by Norwegian botanist Axel Blytt. In his "Essay on the Immigration of Norwegian Flora," he detailed from bogs and sediments how the tundra established itself in the path of the retreating glaciers—later to be replaced by birch, then birch and pine, and later by broadleaf trees.

Corals

Corals provide a good indication of the annual cycles of El Niño in the southern hemisphere. El Niño is important because it is the largest recurring climate event in the world, outside of the seasons. Changes in El Niño weather have the ability to drag along the weather of the entire tropical region. Furthermore, comparisons with prehistoric El Niño records recovered from corals suggest a systematic relationship between global conditions and the character and strength of El Niños.

Sea Bottom

We can learn a number of things from the bottom of the sea. For example, marine plants incorporate stiffer molecules in their cell walls to offset the softening effects of higher temperatures, and the fraction of stiffer molecules in sediments yields an estimate of past temperatures.

Scientists at the University of California at Santa Barbara were able to put together a record of the climate history of the earth going back over the last five ice ages from the ocean bottom in the tropical Pacific. They discovered that forams, a tiny sea creature more formally known as foraminifera, have shells that differ in their magnesium content depending upon the temperature of the water.

Most ocean bed deposits accumulate extremely slowly. One fourth to 1 $\frac{1}{2}$ inches (1 to 4 centimeters) per century is typical. They incorporate not only the remains of marine organisms but also mineral dust blown by the winds. Although they are not detailed indicators of relatively small periods of time, they can yield extremely long records.

Climatoids

The longest sedimentary record so far is from the equatorial Pacific and was analyzed by Cambridge University scientists. It goes back more than two million years and indicates that the ice ages have occurred approximately every 100,000 years during the last million years or more.

Atomic Subs

Another indicator of global climate is found in the Arctic. The Arctic Ocean has warmed markedly in the last two decades, and scientists really don't know why. Knowledge of the Arctic lags behind that of the world's other oceans. Covered by ice year-round, the Arctic is virtually inaccessible to ship traffic and has largely been the province where nuclear subs tried to hide from one another on under-ice missions during the Cold War.

Now, however, some of those subs are taking scientists along for their first sub-Arctic views as part of the Science Ice Expeditions (SCICEX) program, spearheaded by the U.S. Navy and the National Science Foundation. Their discovery is that the ocean's ice cover has thinned by an average of 4 feet (1.2 meters)—some 40 percent—since the 1960s. Some scientists think that the ocean's summer ice-over could completely stop at some point in the coming decades. The Arctic Ocean may have lost its summer ice 400,000 years ago when the earth was as warm as it is now.

Antarctic Ice Cores

The British Antarctic Survey has accumulated ice core evidence that presently goes back 650,000 years. Their core was taken from Dome C, high on the East Antarctic plateau. The results confirm that during this time period, the Earth has experienced eight ice ages, when the climate was much colder than it is today, and eight interglacial periods when the earth's temperature was similar to today. There are many more ice cores in Antarctica, though Dome C is by far the longest one now.

In records that cover the last 650,000 years, it appears that Earth has undergone natural changes from glacial conditions to warmer times like the present. When temperatures are warm, the CO_2 concentration is high—and when temperatures are cold, the CO_2 concentration is low. By comparing the past climate with global environmental conditions today, British Antarctic Survey scientists conclude that we should expect the present warm period to last another 15,000 years. Global warming could start cooking us relatively soon, but we're not headed for another ice age—at least for a while. However, current levels of CO_2 and methane in the atmosphere are higher now than at any time in the past 650,000 years.

The Least You Need to Know

◆ We've been taking the temperature of some parts of the world since Galileo's existence.

◆ To get a picture of the climate before that requires us to look at glacial ice, tree rings, sea bottoms, corals, and lake sediments.

◆ Scientists taking ice core samples in Greenland have a 110,000-year record of the climate, including direct samples of ancient air.

◆ Scientists taking ice core samples from Antarctica have a record that goes back 650,000 years.

◆ All evidence leads to the conclusion that the world is getting warmer and carbon dioxide concentrations are increasing in the atmosphere.

Chapter 4

The Universe Is a Gas

In This Chapter

- ◆ Hydrogen got it started
- ◆ Evolution of the elements in the stars
- ◆ The atmospheres of the solar system's planets
- ◆ How our current atmosphere evolved

The earth's atmosphere is very thin. If the earth were the size of your average globe, the atmosphere would be as thick as a sheet of paper or perhaps a coat of varnish. But it's where we all live and breathe. Without it, we wouldn't be doing much of either. To understand the delicate balance of the atmosphere and how it relates to weather and global warming, let's take a look at how it all got started. In this chapter, I'll discuss the evolution of atmospheric gas in the universe and on Earth from the big bang to global warming.

First Thoughts

Not only has our atmosphere evolved—but our perception of it has, too. When ancient astronomers looked out into space, they thought they were looking at the inside of a big dome (like the Astrodome, only bigger). Their idea was that Earth consisted of Europe, Asia, and Africa surrounded by a huge circular body of water called the River Ocean. Anaximenes, an early Greek philosopher (d. 528 B.C.E.), projected that everything was made of air—only in different densities. When it was thin, it was fire. When it condensed, it became wind and clouds; when it became more condensed, it became water, earth, and stone. Aristotle (384 to 322 B.C.E.), like most other philosophers of his time, felt the world was made of air, fire, earth, and water. His model of the universe showed the planets and the stars rotating around the earth.

It was a long time before Copernicus in the 1500s put the idea of the earth as the center of the universe to rest when he described the planets revolving around the sun—not around the earth. Copernicus argued that the motion we see in the heavens is the result of the earth's daily rotation on its axis and yearly revolution around the sun.

Bang

It took until the beginning of the twentieth century for astronomers to really get a fix on how big the universe was and what it was made of: mostly gas with lots of space between gas particles. U.S. astronomer Edwin Hubble was the first to discover other galaxies. Not only were we not at the center of our galaxy, but our galaxy was also only one of billions of galaxies. He also figured out that they were moving away from us … and each other. In other words, the universe was expanding. Hence, scientists concluded that the universe must have very long ago been compact and dense. And the explosion of that primary matter was the birth of the universe—an event that today we call the *big bang*.

def•i•ni•tion

The **big bang** is a theory that the universe began its existence about 13 billion years ago as the explosion of a small, super-dense concentration of matter—transforming it into gas.

I visited the Yerkes Observatory in Williams Bay, Wisconsin, not too long ago and got to view some of these stars with astronomer Kyle Cudworth, director of the observatory. It was in December, and a series of winter storms rippled through the atmosphere above Lake Geneva. Cudworth took advantage of a break in the storms to look through the observatory's great 40-inch Yerkes telescope at the observatory built by Alvan Clark and Sons. At the beginning of the twentieth century, this was the largest telescope in the world.

What I saw through the telescope was M15, a globular star cluster that orbits just outside our galaxy. Do any of these stars have planets? And how many of those planets have atmospheres or problems with global warming?

Astronomers looking through Clark telescopes like this one discovered Pluto, the fifth moon of Jupiter, the spiral arms of our own galaxy, and the fact that other galaxies were moving away from us—the first evidence of an expanding universe. Telescopes have taught us a lot about our universe and the evolution of materials that make up our atmosphere.

First Gas

NASA scientists looking through today's space-based Hubble Telescope at some of the oldest objects in the universe believe the universe is about 13 billion years old. That's when many believe the big bang occurred. Initially, the universe was pure energy—but some of that immediately changed into matter. Most of the original material was hydrogen gas. As it expanded, it began to clump together. Most of the material in the universe remains hydrogen gas, although it is so light that little of it remains in Earth's own atmosphere. Some scientists look at hydrogen as the ultimate alternative fuel because its combustion releases little or no greenhouse gases.

Stars Form

Stars begin their lives as swirling clouds of gas and dust. As these swirls come together, they reach a point of critical density where the gravity of the individual particles becomes strong enough to draw the elements together. As the cloud comes together, it shrinks up and gets denser and hotter at the core until the region is super hot. At this point, nuclear fusion occurs as hydrogen atoms collide with one another—fusing into helium atoms.

Newborn stars gradually settle down into their adult phase, cranking out energy in the form of heat and light from this conversion of hydrogen to helium. These two gaseous elements are the dominant elements in most first-generation stars. After a star has used up its hydrogen, it will begin to convert helium to heavier elements like oxygen and carbon. These building blocks will be used much later to create carbon dioxide and other greenhouse gases.

Most stars get their energy through the process of nuclear fusion. Four hydrogen atoms come together to create a single helium atom—a reaction so intense it appears as fire. Our own sun converts about five million tons of matter into energy each second. So far, it has used about half of its available hydrogen fuel. At that rate, it will take another 4.6 billion years to run out of gas.

When a star has exhausted all of its fuel, it expands and may become a red giant or a super giant. Eventually, some of these massive stars become supernovas and explode.

In these explosions, the heavier elements are made (including uranium). These elements are thrown back into the universe and again become part of the gas and dust, which may lead to second-generation stars. Astronomers see very few first-generation stars anymore. Most are second to tenth generation. The exploding supernovas create the stardust or elements of which life on our planet is made.

Climatoids

In the song "Woodstock" written by Joni Mitchell and performed by Crosby, Stills, Nash & Young, a lyric claimed, "We are stardust." Whether the words were a mystical notion or scientific inspiration is uncertain, but the idea is correct. All the elements in our bodies and in the atmosphere were created within the fiery furnaces of the stars.

Our Solar System

Our solar system was born about 4.6 billion years ago when a cloud of stardust contracted—perhaps the effect of a nearby supernova explosion. The supernova explosion enriched the cloud with various elements, including silicon, iron, and carbon. As those elements came together, they began to rotate faster—getting hotter in the process. Most of this mass gravitated toward the center of that swirling cloud and congealed into our sun. The gas and dust outside the sun formed the planets, which orbited around it.

There was still plenty of debris in this solar orbit, but over time much of it collided with the new planetary bodies and became part of their masses. Planets like Mercury or satellites like our moon, which lack an atmosphere to buffer those impact points, bear colossal craters on their surfaces to attest to these collisions.

About a billion years of this kind of treatment resulted in the major planets of our solar system. In order of appearance from the sun, they include Mercury, Venus, Earth, Mars, Jupiter, Saturn, Uranus, Neptune, and Pluto. (Pluto, however, was demoted from planetary status in August 2006 by the International Astronomical Union.) Seven of the major planets have satellites; some of the satellites are themselves planet size. Plus, there are lots of asteroids, comets, and meteors out there—the remnants of the original stardust that got us going. None of them have atmospheres.

There are two types of planets in our solar system: *terrestrial* and *gas giants.* The gas giants are massive planets with low densities and thick atmospheres formed mostly from hydrogen and helium. Neptune, Uranus, Saturn, and Jupiter are what we refer to as gas giants. They are thought to have rocky cores, although no one knows for sure. Terrestrial planets are smaller and denser. They are made of rock and have thin atmospheres or no atmosphere at all. Mars, Earth, Venus, Mercury, and Pluto are terrestrial planets. To understand our planet's global warming problem, we need to understand what other planets' atmospheres are like, as well.

def•i•ni•tion

Terrestrial planets are the smaller, rocky planets—Earth, Mars, Mercury, and Venus—which may or may not have thin atmospheres. **Gas giants** are the huge gaseous planets—Jupiter, Saturn, Uranus, and Neptune—which have incredibly thick atmospheres.

Here Comes the Sun

The sun is certainly the most conspicuous body in our solar system. Without its energy, there would be no life on Earth. Most ancient religions have a powerful sun god, and the winter solstice was the scariest celestial event in many primitive cultures. That's because when the sun sank to its lowest point in the southern sky, sky watchers and shamans feared it wouldn't return. So far as stars go, the sun has only average size, mass, and brightness. It takes sunlight about eight minutes to reach our planet—a distance of about 93 million miles (150 million kilometers). The next nearest star is Proxima Centauri, and its light takes about four years to reach us. The sun is about 4.6 billion years old. That's somewhere around midlife crisis for stars. Most of the mass of the sun is a huge gaseous wrap around its core, running at a temperature of about 5,500 K (Kelvin scale)—turning the sun's surface yellow-white. It's not the kind of atmosphere you'd want to raise your kids in.

The surface of the sun is a patchwork of turbulent cells, each about the size of Texas, that bring up energy from the core to the surface. The surface of the sun is anything but placid, with huge loops of gas called *prominences* towering above it. There are also violent eruptions that are called *solar flares*. These events spew light and charged particles into the air. Those particles can cause *auroras*, bad TV reception, and even short circuits in your local power grid.

Strong magnetic fields on the sun's surface can create dark areas known as *sunspots*. Sunspots are dark because they are relatively cool, but the solar output increases with sunspots because the areas around the sunspots are producing more energy. Sunspot activity is cyclical, with peaks occurring every 11 years. Between the years 1645 and 1715 C.E. during the peak of the Little Ice Age, known for its consistent cold weather, there were almost no sunspots. Currently, we are in a period of high sunspot activity. Some scientists wonder about the relationship of sunspots to our present global warming.

def•i•ni•tion

Prominences are huge loops of gases that rise from the surface of the sun and extend far out into space. **Solar flares** are violent eruptions of matter and energy on the sun that spew light and charged atomic particles into space. The **auroras** (known most commonly as the aurora borealis and the aurora australis)—northern and southern lights—are vivid nighttime displays of colored light created when charged particles from solar flares strike the earth's atmosphere near the magnetic poles. **Sunspots** are strong magnetic fields that produce dark areas on the sun's surface.

Mars

Mars is known as the "Red Planet" because of its rust-red hue caused by iron oxide, or plain old rust. It is our closest planetary neighbor and has been the focus of many of our unmanned planetary explorations in recent years. It may once have had an atmosphere similar to ours. Scientists often wonder whether life, even in its simplest forms, ever evolved on Mars.

Climatoids

The idea that there might be Martians running around on the Red Planet was part of the popular culture in the early twentieth century. So much so that widespread panic greeted Orson Welles's 1938 radio broadcast of "War of the Worlds," a fictional account of a Martian attack (based on the science-fiction novel by H.G. Wells).

In the late nineteenth century, Italian astronomer Giovanni Virginio Schiaparelli described what he called *canali* on Mars—what some people speculated might have been canals built by a dying Martian civilization to ship water around the planet. Percival Lowell, a wealthy industrialist, mathematician, and amateur astronomer, built an observatory in Flagstaff, Arizona, in 1894 and spent numerous nights observing and describing those canals. However, photographs taken from exploratory spacecraft sent to Mars revealed no canals or anything that even looked like canals. Many feel that what Lowell really proved was that if you stay up long enough staring through a telescope, you might end up seeing things.

The idea that life may have at one time existed on Mars is not complete science fiction. The atmosphere of Mars is currently only about 0.7 percent as thick as Earth's, but scientists think they might have observed various remnants of an ancient atmosphere in the space surrounding the planet. It's possible that primitive forms of life existed on the planet about three or four billion years ago when Mars may have had an atmosphere similar to ours and was possibly much wetter. But Mars is smaller, has less gravity to hold on to its atmosphere, and has less volcanic activity to recycle atmospheric gases locked in the soil. The CO_2 in its atmosphere provides a small greenhouse effect but is only sufficient enough to raise the surface temperature by about 9°F (5°C). There is simply not enough gas or water vapor in the atmosphere to heat up the planet.

Mars has seasons just like Earth. Mars has polar icecaps, but they are a combination of water ice and dry ice, or frozen carbon dioxide. The changing of the seasons increases the prevalence of dust storms as well as the size of the polar caps. The climate ranges from a high of 60°F (15.5°C)—at noon on a hot day at the equator—to a low of –225°F (204°C below zero).

Jupiter

Jupiter is the largest of the planets in our solar system. It's also a whirling dervish, spinning once on its axis every 10 hours. Jupiter has 16 satellites. The four most prominent are Europa, Io, Ganymede, and Callistro, and scientists are particularly interested in both Io and Europa. Europa has a smooth surface that is probably watery ice. It could be the most hospitable to life in the Jupiter system. Io is also interesting because it has some of the most magnificent volcanoes in our entire solar system. Magma, mainly molten sulfur, is thrown hundreds of miles into space—far above its thin atmosphere. The low gravity of Io may enable the highflying fireworks displays.

Jupiter itself has an atmosphere mostly of hydrogen and helium, the two basic elements of the universe. The brightly colored surface of Jupiter comes from trace elements of methane and ammonia. Storms and atmospheric disturbances are common on Jupiter's surface. One such storm, The Great Red Spot, stretches over the surface of Jupiter in an area larger than Earth and has been blowing for more than 300 years.

Venus

If you want to get the feel of what a runaway greenhouse effect might give you, you only have to look as far as Venus. Venus is the third-brightest object in the sky. It's closer to the size and gravity of Earth, so it retains its atmosphere. The Greeks associated the planet with Aphrodite, the goddess of love. Venus is the goddess's Roman name. Each planet has a symbol, and Venus's is commonly used as the symbol for the female sex. The International Astronomical Union is considering naming the craters on Venus after famous women. Gertrude Stein, Pearl Buck, and Margaret Mead are current candidates.

The atmosphere on Venus is very thick compared to Earth, although not as thick as Jupiter. Venus has an atmospheric pressure that is 90 times what you experience on Earth's surface. You can't see the surface of Venus because it is covered by clouds. The clouds are mostly sulfuric acid, and the atmosphere is mostly carbon dioxide (about 96 percent) with a little nitrogen. What does that get you? A runaway greenhouse effect. At the top of those clouds, it's below freezing—but on the surface of Venus, it's more than 900°F (500°C). Now that's what we call global warming.

The greenhouse effect causes only a mild 60°F (33°C) rise in the surface temperature of the earth, but it raises Venus's surface temperature by almost 900°F (500°C). Although Venus is closer to the sun, that's not the reason why it's so hot. The clouds reflect 75 percent of the sun's rays. However, the greenhouse effect on Venus is so efficient that it prevents all but a negligible portion of the heat generated at the surface from escaping into space.

Earth

Earth is the prototypical terrestrial planet: small, rocky, and possessing a thin atmosphere (as opposed to the voluminous atmospheres of the gas giants). Earth is a layered planet. It has a solid core at the center, a molten core surrounding that, a mantle, and a crust upon which the landmasses and the oceans rest. Scientists believe that the crust of the earth is divided into various plates, which drift in relationship to each other. When these plates bang up against each other, they can create volcanoes that spew carbon dioxide and other elements into the atmosphere.

Earth's atmosphere consists mostly of oxygen and nitrogen. But this stable mixture didn't always exist. Our current atmosphere is the result of a very gradual evolution that began soon after Earth was born, about 4.6 billion years ago.

Volcanic Earth Spews Carbon Dioxide

The first atmosphere—the one that was around shortly after the planet formed—was probably swept away by solar winds, vast streams of particles emitted by the sun. But as the earth cooled, a solid crust was formed and the gases dissolved in the molten rock were gradually released. The atmosphere was then very similar to what is spewed into the air by volcanoes. The principle components of this new atmosphere were probably water vapor, carbon dioxide, and nitrogen.

As the earth continued to cool, the clouds formed and great rains began to fall. At first, the rain evaporated before hitting the surface of the earth. But the process accelerated the cooling, and eventually the rain reached Earth and began to fill the oceans, rivers, and lakes. This took a lot of the water vapor from the air, and a fair amount of carbon dioxide as well. However, carbon dioxide was still the dominant component in the air.

But how did the oxygen-rich atmosphere that we now have come into existence? The volcanic processes that existed on the early planet didn't create oxygen. Some water vapor (remember, H_2O) is carried into the upper atmosphere where the sun's ultraviolet radiation separates it into hydrogen and molecular oxygen (O_2)—but this is a very slow process and can't account for all the oxygen we see in the atmosphere today.

Life Changes It to Oxygen

The main source of oxygen, it turns out, is life itself in the form of green plants. By the process of photosynthesis, plants use sunlight to change water and atmospheric carbon dioxide into organic matter—and in the process release oxygen back into the atmosphere. Scientists believe that the first form of life on Earth may have been bacteria that carried out their metabolism in the absence of oxygen. Today, some of those *anaerobic* bacteria still exist, although most animal life—birds, insects, reptiles, and mammals—are *aerobic*, requiring oxygen to metabolize.

Scientists estimate that the earliest photosynthesizing microbes bloomed in the sea some 2.8 to 3.5 billion years ago. At that time, the atmosphere contained perhaps 1,000 times more carbon dioxide than it does today. But it was the evolution of plants that supplied most of the oxygen that eventually led to our current atmosphere, which

def•i•ni•tion _____

> **Anaerobic** means living or occurring without oxygen. **Aerobic** means living or occurring in the presence of oxygen.

supports the more complex forms of life. With plants giving off oxygen, the percentage of oxygen in the atmosphere slowly increased. The fossil record indicates that by the beginning of the Paleozoic Era some 570 million years ago, oxygen-requiring organisms were abundant in the sea. From a primordial envelope of nitrogen and carbon dioxide, the atmosphere had changed to a mixture of about 78 percent nitrogen, 21 percent oxygen, and a trace of carbon dioxide and other gases. Two million years later, during the Devonian Period, land plants became widespread.

Man Changes It Back?

The earth's atmosphere has evolved over time from one that was free of oxygen to one that has significant levels of free oxygen. In the process, a large portion of the carbon dioxide that was originally mixed into the atmosphere during volcanic eruptions has disappeared. The question is, are we now returning the atmosphere to one where carbon dioxide will again play a dominant role—as it did before the evolution of more complex animals? Right now, man is engaged in a colossal experiment—dumping tens of billions of tons of carbon dioxide and other heat-trapping gasses into the air. Humans have a history of adapting to shifts in the climate. But the fact is, the shift we are now facing might come more rapidly than any we've experienced before.

The Least You Need to Know

- The original gases that made up most of the primitive universe were hydrogen and helium.

- Nuclear reactions within the stars and tremendous explosions at the end of a star's life convert simple gases to more complex elements.

- We all are made of stardust.

- The solar system is comprised of both rocky and gaseous planets.

- The atmosphere on Earth is the evolutionary product of volcanic gas (CO_2) converted to oxygen, primarily through the photosynthesis of plants.

Part 2

A Historical Perspective

So now you've discovered that greenhouse gases are increasing in the atmosphere, and it looks like our machines are putting them there. What do we do next? To answer that question, scientists have had to look into the past to see how these gases have varied over geological time and what the consequences were for the atmosphere.

Ocean bottoms, ancient trees, and polar ice give us a glimpse into the past of the earth's climate. What we see is that this modern era of climate stability may be a false prophet. The atmosphere is really a roller coaster, and wide swings in temperature are the norm rather than the exception.

I'll take a look at how the climate affected the Mayans and Vikings, and what the Little Ice Age that peaked in the early 1700s did to us all. Then I'll talk about how industrialization and deforestation are affecting our atmosphere right now.

"I'm sure I just heard it say, 'Back in my day.'"

Chapter 5

Climate Is a Roller Coaster

In This Chapter

- ♦ The climate can change abruptly
- ♦ Changes in the last ice age
- ♦ Changes closer to the present
- ♦ Breaching climate thresholds

In the previous chapter, I talked about how scientists have collected evidence of the climate from the past—digging in ice, coring around in old lakebeds, and bringing up samples from the very depths of the sea. Much of the new evidence has only been gathered since the International Geophysical Year of 1957–1958. But most of the big push has occurred in the last two decades as scientists addressing the problem of global warming look to the past to see what it can tell us about the future.

In looking at the past, scientists have paid particular attention to abrupt changes in climate and how they have affected Earth. Society is used to gradual changes. But what would happen if different climate conditions came suddenly? What if you woke up one winter day in New York City and the weather was as warm as Miami, Florida's? And what if it stayed that way? How would that affect life, the environment, and the economy?

We're Spoiled

We take our fair weather for granted. The rain we have, the storms we get, and the droughts we endure seem severe. One summer is cool; the next is hot. One winter it snows a lot; the next there's hardly any snow at all. The weather is wild. It must be changing. Didn't I hear something about global warming? Is that what's going on here?

It's true we are under the influence of global warming, but perhaps the reason we're so complacent about it is because we're spoiled. The weather we've endured—that flood where your car got stuck, the fire that stopped your vacation, the hurricane that blew the windows out of your Aunt Myrtle's summer home in Florida—is, I'm afraid, peanuts compared to the climate records that scientists have uncovered from the past.

The earth right now is in an *interglacial*. It's the period between ice ages (or more properly, glacial episodes). Ice ages are a series of glacial episodes. There have been at least four major ice ages in the Earth's approximately 4.5 billion-year history. Outside that, the earth has been relatively ice-free. In our current interglacial, the world is warmer. Those glaciers that once hung outside Chicago and New York are gone. When I step out my door in southern California, the climate isn't like Vancouver one year and Acapulco the next. But according to ice cores, tree rings, and sediments, such climate swings may once have been the norm.

def•i•ni•tion

Interglacial refers to the period of relatively stable climate between glacial episodes or ice ages. The earth is currently in an interglacial.

The Ice Ages

The longest continuous sedimentary record to date goes back more than two million years and indicates that glacial episodes have occurred approximately every 100,000 years during the two million-year-old ice age we are in right now. To get an idea of what it was like before that, scientists look in Earth strata for deposits of rock and clay that tell them there's been a glacier there. From this evidence, scientists believe there have been at least four major ice ages in the past—periods when ice covered extensive areas of the earth. The last ice age was from 270 to 350 million years ago. It may have had more than 30 glacial epochs.

Prior to the current ice age, we were in a hothouse climate where there was little or no permanent polar ice. Presently, about 10 percent of the earth's water is locked up in ice—mostly in the Greenland and Antarctic ice sheets. About 20,000 years ago, approximately 30 percent of the world's water was locked up in ice, and Greenland temperatures were about 20°F (11°C) cooler. The coldest parts of the ice age might have been as much as 40°F (22°C) cooler.

Peaks and Valleys

It's hard to talk about average temperatures when describing the most recent glacial episode. The climate didn't just get cold and stay that way; it swung wildly back and forth. Over the last 10,000 years or so, the temperature has stayed relatively warm and constant. But about 12,800 years ago when the earth was coming out of the last glacial episode and things were warming up nicely, the earth suddenly took a big dive back into the cold. Then, a little later it jumped back into the warmer period. It was called the Younger Dryas (YD) event.

The Younger Dryas event might have been just business as usual as far as the climate was concerned. The Greenland record shows that during the last glacial episode, large abrupt climate changes occurred 24 times. Evidence from the glaciers, salt from the ocean, methane from the tropics, windblown dust from Asia, and snow from Greenland also indicate that those big climate swings took place worldwide.

Over the last 100,000 years, there have only been two vaguely stable periods in the world's climate. One was a short period back in the middle of the last ice age when the weather was at its coldest. Central Greenland was 40°F (22°C) colder than it is today. (Imagine what it would be like if the average temperature of the place you call home dropped that much.) The other period of stability is the last 10,000 years. The rest of the time, it's a climate roller coaster. Some of these dives and jumps took less than a decade—perhaps a few years.

Global Warnings

*Some say the world will
    end in fire,
Some say in ice.
From what I've tasted of desire
I hold with those who favor fire.
But if it had to perish twice,
I think I know enough of hate
To say that for destruction ice
Is also great
And would suffice.*

—Robert Frost, "Fire and Ice"
(1923)

Solar Wobbles and Wiggles

So how do we get the glacial episodes of our present ice age? Why does the world seem to be on a 100,000-year cycle of glacial ice? (Remember, if we can figure out the past, maybe we can figure out the future.) The earth rotates around the sun, but the path it takes is not a perfect circle. The orbit of our planet is out of round; they call it "elliptical" (sort of like a football, only not so pointed). But on a 100,000-year cycle, the orbit changes. The orbit becomes less like a football and more like a basketball, which affects the climate. Note that the 100,000-year cycle matches the length of the glacial episodes.

There is another variable. The earth's axis is on a tilt. As it travels around the sun, it's a little cockeyed like your globe is. This is what makes the seasons. The northern hemisphere of the globe is tipped toward the sun on one side of the orbit, which is when the northern hemisphere has its summer. The northern hemisphere is tipped away from the sun on the other side of the orbit, which is when it has winter. The axis also wobbles over time.

Scientists have long looked to the eccentricity of the orbit and the tilt and wobble of the axis as the main reasons for the glacial episodes in our current ice age. They don't increase or decrease the total amount of sun the planet gets, but they change where and during what season the sun falls. This is enough to cause the glacial episodes—or so we used to think. However, we are now in the age of computer models. We can take certain climate variables, feed the data into a supercomputer, and get a prediction. We can also do it the other way around: feed the result into the computer and see whether the variables we are investigating give us the result that we already know. And the problem is if we feed the eccentricity of the earth's orbit and the tilt and the wobble of its axis into a computer, they are not enough to produce the glacial episodes in the current ice age. Maybe it helped get things going in the right direction, but it didn't pull off the freeze all by itself. So … what did?

Who Done It?

Could it be the ice itself? Ice has a greater *albedo*. It reflects sun more than the forest and more than the dark seas, which absorb the sun better and heat the place up. Ice sends sun packing. If you get more ice, you get more reflection—so it could perpetuate itself. But ice does some funny things that keep it from taking over. As an ice sheet grows bigger, it acts as a blanket, increasing the pressure and warming up the earth beneath it. Plus, the ice sheet is so heavy it sinks the earth beneath it to altitudes low enough to raise the temperature. When it gets warm down there, the ice can lose its frozen grip on the land and slide out to sea. Or it can simply drown in a pool of warm water. In other words, glacial ice normally has a few governors that keep it from becoming self-perpetuating.

def•i•ni•tion

The **albedo** of a surface is its ability to reflect back the light that falls upon it.

So what else? Well, how about carbon dioxide? As it turns out, the same stuff we're worried is going to make us warmer also had a major role in making Earth a lot colder. Carbon dioxide in the atmosphere, deep-sea temperatures, and Earth's orbital eccentricity all vary in step with the same 100,000-year cycle as the present glacial episodes or ice ages. Ice volume lags behind a bit, apparently ruling it out as a prime mover. Deep-sea temperatures may have something to do with the conveyor belt–like currents that are operative over most of the oceans, moving warm water and cold water at different levels. A breakdown in this system may be a part of global warming.

But then, there's our old friend carbon dioxide. For reasons that scientists are still not clear about, it seems that carbon dioxide decreases with the beginning of the ice ages and increases on the way out. Scientists have seen this phenomenon in the Greenland ice cores but also in the much longer records of ice cores in the Antarctic, which go back over 650,000 years.

Richard Alley feels that, for reasons we don't understand completely, the cooling of the ice-age cycles driven by the eccentricity of the earth's orbit reduces the greenhouse gases, particularly carbon dioxide and water vapor. Water vapor fell when ice-age cooling reduced evaporation. Carbon dioxide fell, he believes, when blustery ice-age winds blew more dust into the oceans—fertilizing the algae. Algae (or phytoplankton) sucks up the carbon dioxide; animals eat the algae; and the carbon dioxide ends up as old poop or old shells at the bottom of the sea. Colder seas also suck up more carbon dioxide even without the little plants and animals.

Snowball Earth

Sure, the last glacial period of our present ice age was rough. But the saber-toothed tigers, wooly mammoths, and ice-age man—who used to complain about it so much—didn't know how good they had it. What if they'd been around during snowball Earth? The present ice age is peanuts compared to what went on 600 million years ago during a period known as the Neoproterozoic, when some scientists believe that the surface of the planet almost froze solid.

Scientists got their first hint of snowball Earth from glacial debris found near sea level in the tropics. The thing is, most glaciers near the equator never get down below 16,400 feet (5,000 meters). Even in the coldest winter of the last glacial period, they never went lower than 13,100 feet (4,000 meters). If scientists who studied snowball Earth can be believed, the worldwide global temperature got down to an average of −58°F (−50°C). Heat, ultimately generated by the hot core at the center of the earth, kept the ocean from freezing to the bottom—but it got so cold that sea ice froze to six tenths of a mile (1 kilometer) in thickness. Microscopic life on the planet was nearly wiped out.

Scientists believe that torrential rain (which binds to carbon dioxide) removed carbon dioxide from the atmosphere during a period when most of the continents were gathered around the tropics and it rained a lot. This and the albedo of polar ice caps led to a runaway freeze.

Hothouse Earth

But then what? Well, after snowball Earth we had hothouse Earth. Although the earth was frozen, volcanoes spewed a steady stream of carbon dioxide into the atmosphere. Carbon dioxide levels increased about 1,000-fold over 10 million years of normal volcanic activity. The ice melted, and the runaway hothouse formed. In a matter of a few centuries, a brutally hot world replaced the former deep freeze. Surface temperatures soared to more than 122°F (50°C), driving an intense cycle of evaporation and rainfall.

Only a tiny fraction of life survived snowball Earth. In terms of biological destruction, it was more deadly than the period following the asteroid hit that took the dinosaurs out 65 million years ago. Then came hothouse Earth, and the scorching heat further thinned the small numbers of organisms that survived the ice. But then, Earth settled back to normal. A rapid diversification of multicellular life followed, resulting in the *Cambrian explosion* of organisms between 575 and 525 million years ago. It is the time when most of the major groups of animals first appear in the fossil record. It seems that hard times promote biodiversity.

def•i•ni•tion

The **Cambrian explosion** or Cambrian Period marks an important point in the history of life on earth. It is the time when most of the major groups of animals first appear in the fossil record. It is often referred to as the "Cambrian explosion," because of the relatively short time over which this tremendous diversity occurred.

What's impressive in many of these great catastrophic swings in the climate is the role that carbon dioxide plays. It is one of the main drivers of the climate on Earth. Swings in its concentration have molded the land and left massive scars on the terrain as well as on life itself. The more scientists study its effects, the more they realize it's not something to play around with.

Meet the Younger Dryas

But, you say, this was all a long time ago. Most of this stuff took eons. The geological history of our planet is on a clock that moves over such vast amounts of time that it's not applicable to human life. Why worry about something that moves so slowly that it can't possibly affect you in your lifetime or your children's lifetimes—or even your grandchildren's lifetimes?

Well, let me introduce you to the Younger Dryas. The event occurred about 12,800 years ago during a period of increasing warmth, when the world was suddenly and very drastically thrown back into the last glacial period of the present ice age. It remained that way for more than a millennium, until about 11,500 years ago, when it swung back out. The average Greenland temperatures at the end of the Younger Dryas event jumped 15°F (8.4°C).

The data indicates that the cooling into the Younger Dryas event occurred in a few decade-long steps, whereas the warming at the end of this enormous climate swing occurred primarily in one especially large step of about 14°F (8°C) in around 10 years. That jump was accompanied by a doubling of snow accumulation in three years, most of the change occurring in one year.

Scientists digging into peat bog sediments in Europe in the late 1800s were the first to discover it. The *dryas* is an arctic flower in the rose family that is white with a yellow center. It's found in cold tundra or in high altitudes. Digging through sediments to determine the sequence of plant types following the retreat of the glaciers, scientists found a number of layers representing different forest types that occurred as the earth warmed up. But then, they found the small little fossil leaves of the dryas.

Going deeper, they encountered three layers of these plant sediments before encountering the rock and gravel that were laid down by the glaciers themselves. Those layers represented the Younger Dryas, the Older Dryas, and the Oldest Dryas. The Younger Dryas was the most severe. It's also the most studied of any abrupt climate change on our planet.

def•i•ni•tion

The **dryas** is an Arctic white flower with a yellow center—a member of the rose family—found in the tundra. It's used as an indicator of Arctic conditions in sedimentary layers.

Earth's Climate vs. the Younger Dryas

Ice core evidence from Greenland during the Younger Dryas shows that windblown dust and sea salt increased by a factor of three to seven times. In Norway, mean July temperatures were about 12 to 16°F (7 to 9°C) colder than today. In Spain, July temperatures were as much as 14°F (8°C) colder than today. For a while, scientists thought the event was mostly confined to Europe as the result of a shutdown in deep North Atlantic currents. But Dorothy Peteet, a senior research scientist at NASA in New York, found evidence of the Younger Dryas in lakebed sediments in the Adirondack Mountains and on the islands off Alaska.

She found that the tundra moved all the way to Virginia, replacing the forests. When she'd gotten tired of looking at the Younger Dryas in the Northeast, she went on an expedition to Alaska where she discovered lakebed sediments of fern overlaid by tundra plants on Kodiak Island. "We joked with each other that it was probably the Younger Dryas, and it turned out that was exactly what it was," she said.

Younger Dryas evidence also showed up in Central America, the Caribbean, and South America. Marine sediment cores from the event show an increased abundance of polar plankton. This finding hints that deep ocean currents might have weakened or shut down during the Younger Dryas—and may even be the reason for it.

But the greatest catastrophe of the Younger Dryas event according to Dorothy Peteet was the extinction of the great animals of the present ice age. The mastodons, wooly mammoths, horses, and saber-toothed tigers that once wandered the North American continent became extinct during the Younger Dryas.

Hot Debates

Scientists debate whether the extinction of the mastodons, mammoths, horses, and saber-toothed tigers—which occurred during the Younger Dryas (12,800 to 11,500 years ago)—was the result of early man's invasion of North America or the massive swing in climate during that period. Many scientists believe that both of these causes contributed to one of the highest periods of mass extinction in North America.

What If It Happened Today?

According to Peteet, if an event on the scale of the Younger Dryas warming (the period coming out of its cooling) was to occur in the Northeast, it would be catastrophic. "The weather in New York and New England would suddenly become like Florida. We would lose all the trees in the Adirondack Mountains—the spruce, fir, and paper birch. We'd also lose the oaks and white pine. The dead wood would create major forest fires. And trees from Florida wouldn't move up to replace them. We don't have migratory pathways. We have highways and agriculture instead," she said.

Rapid changes such as the Younger Dryas are basically harder for humanity and nature to endure. While the economy, industry, and humanity may be able to replace, repair, and adapt to changes that occur over time, rapid change is tough. A rapid sea level rise, for example, would devastate coastal buildings and harbors.

Rapid warming could destroy forests, vineyards, and fruit trees—not to mention the ski industry. It could rapidly deplete the groundwater reserves of the Southwest desert. You could expect your property value to fall if your place does not have proper insulation or cooling. Droughts would devastate native trees, which need water to fight off insects.

The Mechanics of Abrupt Change

So what do we mean by an abrupt climate change? Twenty-first century humanity thinks of the climate as something that, although it may vary from year to year, stays pretty constant. You can rely on the Northeast for cold, the Northwest for rain, the Southwest for sun, and the Southeast for subtropical storms and sun. But that's not really the way the climate works. That's just our recent experience. The way the climate really works on our planet is that it jumps all over the place.

The analogy of abrupt climate change is shown in the following figure. Imagine a balance consisting of a curved track where there are two cups in which a ball might rest. The system has three points of equilibrium: (a), (b), and (c). The middle point (b) is unstable. If the ball is moved ever so slightly to the left or right, the ball will move to a point far from its original position. However, if you were to move (a) or (c) a bit, the ball would merely rock back and forth and then settle back into the (a) or (b) cup. But if you push down on either (a) or (c) gradually—more and more—eventually the ball will roll over the fulcrum and pass into the opposite cup. This is an example of a system passing a threshold. When the pressure is relieved, the system does not return to its former state. It has entered a new regime.

Analogy of abrupt climate change.

(Photo courtesy of National Academy Press)

This is what scientists feel happened with the Younger Dryas. There was steady force exhibited by various climate features until the climate shifted into an entirely new regime. The climate over most of the world changed suddenly and drastically.

A good example of a system passing a threshold and entering a new regime is the floods on the Red River in Grand Rapids, North Dakota, in the late 1990s. Since the nineteenth century, Grand Rapids had successfully fought frequent floods on the Red River to a river stage of 49 feet. Then, in 1997, the water level crested the levees at 54 feet and led to a catastrophic flood. You might say that the threshold had been passed, after which the levees failed. When the city rebuilt the levees, planners considered past events to determine what was the highest that the river might rise.

Another example is the massive failure of levees in New Orleans, Louisiana, in the face of Hurricane Katrina. The U.S. Army Corps of Engineers issued a 6,113-page report on that failure. Essentially, the report claimed that the system was built piecemeal over 40 years and the design was based on outdated scientific data. It was designed to with-stand relative weak hurricanes, not Katrina-size monsters. Also, engineers had failed to take into account how much Louisiana soil had sunk in recent years. In some cases, levees were two feet lower than they were meant to be. The pumping system that was supposed to mitigate post-hurricane flooding was quickly overwhelmed. Katrina exceeded the threshold of the levees and flooded 80 percent of the city.

The trouble is, the world might be moving into a new era where thresholds for individual systems like the levees of the Red River and New Orleans as well as large systems like the deep North Atlantic currents may well be breached. The constant pressure that we are putting on this system by a gradual rise in carbon dioxide may lead to a corresponding gradual rise in temperature—or it may lead to a sudden, wide-spread, and catastrophic change.

And it just might stay that way for 1,000 years.

The Least You Need to Know

- The climate over the last 100,000 years can best be described as a roller coaster with lots of steep drops and rises.

- We are currently in a period of calm following the last ice age or glacial epoch.

- The current period is not typical of the earth's climate. The climate in the past has swung from a solid ball of ice to a scorching hothouse.

- The Younger Dryas event was a dramatic climate swing in and out of a cold spell that occurred from 12,800 to 11,500 years ago.

- Our climate is like a balance in which the ball could roll from one side of the ful-crum to the other and not go back for 1,000 years.

Chapter 6

More Recent Changes

In This Chapter

- Ice-age hunters start growing stuff
- The weather and early civilizations
- A four-century dip back into the cold
- The Dust Bowl

In the previous chapter, we described the harsh environment that existed before the Holocene—the relatively warm and stable 10,000-year period we are living in today. Although modern man was around for at least 30,000 years before the end of the last ice-age cycle, it was only when the climate settled down and things got warmer, more stable, and less stormy that civilizations started to flourish.

Early Man Warms Up

The earliest historical records that ancient humans left us are from sketches and paintings on the walls of caves in France and Spain, which scientists date back 15,000 years. Those pictures show a world of horses, deer, wild cattle, rhinoceroses, and mammoths that early hunters chased down with spears and arrows. The pictures depict a treeless landscape where scientists tell us that cold, dry winds, storms, and widely varying temperatures dominated.

Those Burly Hunters

Man populated areas that were generally beyond the range of the great ice sheets. Scientists have uncovered a number of their remains, which show the hardy life these guys endured. An ancient encampment in northeast Russia is filled with the bones of mammoths, which means that they weren't just dreaming about the big beasts—they were actually living on them.

Archeological finds show human beings lived in the ice-free areas of Alaska north of the big ice sheets. Humans probably roamed back and forth from Siberia to Alaska across the dry plain that then linked those lands.

During the last ice age, the expanded glaciers and ice sheets locked up enough water to lower sea level around the world by about 400 feet (125 meters). This was apparently enough to produce the Bering Land Bridge from Siberia to Alaska, across which man first ventured into the Americas. Similar great stretches of dry land might have also existed between Asia and Australia, providing the first path man took to the land down under.

About 15,000 years ago, the world started to defrost. The glaciers receded, vast sections of the ice caps melted, and the seas began to rise. As the world warmed up, more water was available to the atmosphere and rains increased. The great ice sheets that covered Scandinavia during the last ice age began to melt about 8200 B.C.E. and were about the size they are today around 6000 B.C.E. The sea invaded northeastern Canada's Hudson Bay about 6000 B.C.E., and the North American ice cap—which covered much of Canada—rapidly melted.

The sea level rose about 10 to 16 feet (1 to 5 meters) per century. Although those numbers don't seem high, modern experience with sea level rises on the low-lying coasts of the North Sea shows that the coastal land doesn't give way gradually but in sudden retreats during great storms that arrive as storm surges at low tides.

Settling Down on the Farm

Early man lived in mountain caves close to areas where the game was plentiful. However, about 10000 B.C.E. the human inhabitants in the Zagros Mountains in the Middle East began to change. Understanding their story involves picking through their remains. Up until then, they'd hunted mostly wild sheep and goats. But then sheep became more important to their diet, and the proportion of young sheep bones indicates that they must have started herding.

A large number of grinding tools soon appeared as man began to collect grains. And a little later, wheat and barley were found in the steppes near the mountains in areas where neither the climate nor the terrain was suited to the wild forms of these grains. From this evidence, scientists conclude that humans must have discovered agriculture. It seems the last time the atmosphere warmed up substantially it was good for man. But will the coming warming be a bonus for humans, or a tremendous disaster?

As the world warmed, it got drier. Cave paintings from the Sahara Desert around 3500 B.C.E. show men hunting a hippopotamus from a canoe. Flood records from the Nile River show an increasingly dry world where elephants, giraffes, and rhinoceroses of Egypt disappeared from 3,000 to 2,000 B.C.E.

Climatoids

The rise of civilization can be traced to that moment when early humans left the caves and the more abundant game of the mountains and ventured into open sites in the foothills, which were more conducive to agriculture. This transition took place only after the world began to warm.

The Peak of the Holocene

The Stone Age evolved into the Bronze Age, and the warm weather and freedom from storminess allowed for the development of culture through trade over land and sea. In the southwestern portion of the United States, the first agriculture spread northward with increased subtropical rains in the warm period between 4500 and 4000 B.C.E. Native American societies began to flourish.

Tree ring data taken from the bristlecone pines in the White Mountains of California give us good evidence of the history of warmth and rain. That area has gradually dried up, and desert has taken over what was once trees and grasslands. A series of lakes that once pooled in Owens Valley between the White Mountains and the Sierra Nevada Mountains today are mostly dry.

Religion and Weather

Climate has varied throughout the world. In China, there were severe droughts that occurred around 1200 to 800 B.C.E. and again from 600 to 200 B.C.E. Civilization was on the move at that time, perhaps hunting for wetter lands. The movement might have had something to do with the spread of the great religions. Drought brings friction as society struggles for limited resources. It also brings suffering. Buddha and Confucius offered solutions to human suffering, and their messages traveled rapidly around 500 B.C.E. This was also the time of the great Greek philosophers whose teachings widely influenced European thought and the rise of Christianity. Christ, however, seemed to have been born in a period of climate stability.

Climate and Rome

The climate wasn't a great obstacle to the Mediterranean world in the time of the Romans. Julius Caesar apparently had to wait in the summer of 54 B.C.E. while persistent winds delayed his expedition across the channel to England. Similar weather could be found today.

Global Warnings

Though the plague might be most famous for its European invasion in the fourteenth century, it took a toll in Asia in much more recent times. An outbreak in China began in the 1890s and spread over the next 20 years throughout China and beyond, killing more than 10 million. What outbreaks await a warmer world? Drought would exacerbate the spread of disease.

Late in the first century C.E., Roman emperor Domitian prohibited vineyards in the western and northern provinces of the empire. The climate was gentler at the time, and farmers quickly realized that grapes could be grown in places where they hadn't existed before. It was around this time that vine growing began in England and Germany. The edict was revoked by Probus around 280 C.E.; he obviously liked his vino a little more than Domitian.

Greece and northern Africa were probably helped by the colder climates, and this may have been the reason why Roman agriculture proliferated there in that era. The water table, too, might have been higher then—and desert oases more extensive—as a result of moister climates that existed coming into that era.

The Big Plagues

A series of plague epidemics swept through the empire in late Roman times. In the second century, the plague struck twice in Egypt—reducing the population in various regions by a third. In the middle of that same century, the plague moved from Macedonia and spread throughout much of the empire. The plague struck again in the third century. Worst was the bubonic plague, which arrived from the Middle East in the sixth century. It took 100 million people in all. The drought that was prevalent in those days may have contributed by making hygiene difficult due to the lack of water.

That's different from the incidence of the disease in the fourteenth century, where the plague seemed to spring from warm, moist places in cities. Plague is an infectious fever—a bacterium transmitted by the rat flea. Scientists estimate that the Black Death, as it was known in the fourteenth century, killed more than one fourth of the population of Europe—or 25 million people.

Fall of the Mayans

Meanwhile, on the other side of the Atlantic, the Mayan civilization was flourishing in the tropical rainforest of the Yucatan Peninsula in southeastern Mexico. Although the Mayan civilization has lasted through to this day, it reached its apex (so far as control of the population and growth of its great cities) between 300 to 800 C.E.

Mayan ruins at Tikal, Guatemala.

(Photo courtesy of Michael Tennesen)

The Mayans spread their influence throughout Central America and Mexico, flourishing in southern Mexico, Guatemala, Belize, El Salvador, and Honduras. Massive temples and stone monuments typify the classic period of the Mayans. They had an accurate calendar and even made astronomical observations. They had beautiful multicolored pottery and finely carved statuaries. But they eventually failed, and climate may have played a role in that failure.

The population centered in the lowlands of northern Guatemala. Perhaps their best-known site is at Tikal. The center of Tikal contains more than 300 major structures. Temples I and II dominate the central plaza, standing 144 feet (44 meters) and 128 feet (39 meters) high. The adjacent ruins contain palaces and a ball court. Archeologists feel that about 10,000 people lived in an area around the center of the site while perhaps 40,000 more lived in the surrounding rural zone. It turned out that these dense populations were a setup for disaster—particularly when the climate changed.

Although Tikal National Park in Guatemala is best known for its Mayan ruins, it's also famous as a wildlife sanctuary. On a visit to Tikal a few years ago, I remember having to stop and wait while a troop of 50 or more coatimundis (raccoonlike animals) crossed our jungle path. I remember walking down lush green trails with ferns, bromeliads, and orchids hanging from every bough while spider monkeys crawled through the branches overhead.

Scientists formerly believed that Tikal was a ceremonial center where only priests and acolytes went. Now they believe that Tikal was actually an urban center for trade and commerce. Mayans practiced *slash-and-burn agriculture*, and corn was their principal crop. Classic Mayans practiced a variety of intensive agricultural techniques including ridging, terracing, and raising fields in swampy areas. Slash-and-burn agriculture is an agricultural technique that has a limited life span in tropical climates. The initial cutting and burning of the forest releases nitrogen and other nutrients to the soil, but farmers can only get a limited number of crops out of shallow rainforest soils before the nutrients run out. Some scientists believe the Mayan civilization declined because they exhausted the soils that surrounded their classic cities.

def•i•ni•tion

Slash-and-burn agriculture is an agricultural practice in rainforest regions where the forest is cut down and burned so that nitrogen and other plant nutrients in the trees are released into the soils for crops. The nutrients in the soil last only about three years before farmers have to move on and cut down more forest.

But scientists examining sediment cores from Lake Chichancanab in the Yucatan believe that drought might have brought the Mayan civilization to its knees. These cores represent a 2,600-year-long record of the world the Mayans had to contend with.

Climate may have felled the Mayans at the height of their culture. Evidence shows that drought peppered the 2,600-year record, but the most intense and most prolonged drought ran from about 750 to 900 C.E. The onset of this drought period would have resulted in crop failure, famine, and disease. The increased competition for food may have led to revolts against the ruling classes as well as war between neighboring states for limited resources.

Anasazi ruins.

(Photo courtesy of Michael Tennesen)

The Anasazi

The Anasazi Indians, whose elaborate structures flourished in the southwestern United States, may be a similar case. In what is now New Mexico and Arizona, they built an elaborate complex of roads, irrigation channels, and giant pueblos of stone and masonry. They built 75 towns in Chaco Canyon in New Mexico alone. The great house of Pueblo Bonito, perhaps the largest of the Chaco pueblos, appeared around 1000 C.E. Supported by massive timbers, its adobe walls rose four stories high and encompassed 800 rooms. Around these structures were a number of circular excavations known as *kivas* that were used as ceremonial chambers.

At Mesa Verde in Colorado, they built homes into the sides of cliffs. The largest dwelling at Mesa Verde, called "Cliff Palace," is three stories high with 200 rooms and several towers. It probably housed 250 people—each family living in a one-room house. The Indians climbed up to their cliff houses using finger and toe holes carved into the sides of the cliffs. The difficult ascent was a strategy to keep their enemies out.

def•i•ni•tion

Kivas were the ceremonial chambers that surrounded the towns and cliff dwellings of the Anasazi Indians. Their structure was influenced by an earlier day when these Native Americans built homes of covered pits.

But then, in the late 1200s, the Anasazi abandoned these two sites. Tree ring evidence gathered from the massive beams that were used to build some of the later structures again point to drought as the principal culprit in this enduring mystery. In the late 1200s, little rain fell in the region—and the Anasazi might have abandoned their great cities to flee south to areas around the Rio Grande River (where water was more reliable). Soil erosion, warfare, and disease also might have played a part. All of this was a result of climate change.

Vikings Don't Like Cold

Around 800 C.E., the Vikings ventured out into the Atlantic with their new vessels to pillage and explore. Their mastery of sailing gave them an edge over the Europeans. A warming trend occurred from about 950 to 1250 C.E. in Europe, the North Atlantic, and North America. Records from the Greenland ice cores and the California tree rings confirm this. The Vikings soon spread north into the Arctic, west to Greenland, and ultimately to Newfoundland. They did this in their early days without so much as a compass. But it was a relatively warm period—one free from severe storms and accompanied by a long retreat of the sea ice. At its peak, the Greenland settlement had more than 280 farms and a population of about 3,000.

A book written around 1125 C.E. describes the Viking settlement in Greenland toward the end of the first millennium. Then, a man named Thorkel Farserk—a cousin of Erik the Red (who founded the Greenland Colony) who had no boat at hand—swam out to an offshore island to pick up a full-grown sheep and carry it home to entertain his cousin. English scientists believe that 50°F (10°C) would be about the lowest even an obese person not trained in long-distance swimming could swim the distance recorded. The average temperature in modern times for that water seldom exceeds 43°F (6°C), so it seems the water must have been at least 7°F (4°C) warmer during the time Farserk swam for his sheep.

The warmth did not last. The Viking settlement gradually contracted as the climate cooled into the Little Ice Age that eventually followed. The settlers brought their farm animals into their homes when the winters got too cold. Eventually, they ate those farm animals as Greenland was increasingly locked in ice. In the late 1400s, Pope Alexander VI noted that no ship had put into shore on Greenland in decades and the colony was almost unreachable. The Vikings of Greenland eventually died out, although their Eskimo neighbors—the Thule Inuit—survived the cold just fine.

The Little Ice Age

From about 1400 to 1850 C.E., much of Europe, North America, and other parts of the world experienced what scientists refer to as the Little Ice Age. According to Richard Alley, it was not a stupendous plunge back into the ice age (as was the Younger Dryas event), but it was a taste of the instability that is inherent in the earth's climate. Many regions experienced harsher winters than they'd known before. Beginning in the early 1600s, the Thames River in London began to freeze occasionally. *Frost fairs* were held during winters when the ice was solid enough to support festivities. The fair included a small tent city replete with food, festivities, and games such as ice bowling and ice skating.

The cold winters affected French vineyards, held up traffic on the Dutch canals, and changed the course of wars. It created harsh conditions for the American troops at Valley Forge. It might have been the Achilles' heel for Napoleon on his ill-fated march into Russia.

Storms were frequent and nasty. The North Sea floods produced a tremendous loss of life in 1570. A storm in 1634 permanently changed the coasts of Denmark, Germany, and the Netherlands. The great storm of southern England in 1703 blew down the lighthouse near Plymouth, uprooted enormous trees, and demolished towns and the countryside all across England. Ships were blown upriver or wrecked on the coast or at sea. London sustained millions of dollars worth of damage. The storm took more than 8,000 lives.

def•i•ni•tion

From the beginning of the 1600s to the beginning of the 1800s, during a time when much of the world was enduring the Little Ice Age, **frost fairs** were held on the Thames River when the river froze over. Since then, the world has gotten warmer … and now the Thames seldom freezes.

The biggest difficulty in the Little Ice Age was not so much the cold but the storminess and erratic nature of the colder climate. Deep and sudden chills destroyed harvests. In the summer of 1665 and 1666, in the middle of the worst century of the chill, the weather turned extremely hot. London experienced its last great epidemic of the plague. At the end of the summer of 1666, a great fire burned through the city. During the Little Ice Age, temperatures averaging $3.6°F$ ($2°C$) less than in modern times shortened the growing season in England five weeks on average. Although $3.6°F$ ($2°C$) may not sound like much, it points to what a small shift in the average temperature can do. Compare these figures to the 2.0 to $11.5°F$ (1.1 to $6.4°C$) of warming that scientists predict for the next century.

In North America, eighteenth-century winters impacted the lives of New Englanders far more than winters do today. When the waters froze, boats could not fish or haul goods. Deer that were used for meat died in heavy snowstorms in which snow piled up and made it difficult for the deer to escape wolves or bears. In 1717, four snowstorms blanketed the region with five feet of snow from February 27 to March 7. Winds then whipped the snow into drifts 10 to 25 feet high.

In the winter of 1740–1741, Boston Harbor became such an expanse of thick ice that sleighs carried Sunday worshippers across it from late December through early April. One man even traveled 200 miles over the ice from Cape Cod to New York City.

The Dust Bowl

The climate started to warm up in the 1850s. By the second and third decades of the twentieth century, the climatic warming became evident to everyone. One of the tragedies of the warmer weather was the 1930s phenomenon known as the *Dust Bowl*. The Dust Bowl was an agricultural region that spread over 150,000 square miles (390,000 square kilometers) of the plains states of Kansas, Colorado, New Mexico, Texas, and Oklahoma.

Cattle and other stock had been the principal business in this area until World War I. Grain prices jumped, and farmers plowed up millions of acres of grassland and planted wheat. But then, the Depression hit—followed by a devastating drought from 1933–1939. With the demise of the grasslands, cultivated fields—harvested of their wheat and dry from years without rain—were an easy target for the winds that ripped across the plains. Windblown sands piled up in dunes some 30 feet (9 meters) high. They blocked roads and choked pasture grasses. Lighter soils were blown into dust clouds that were 5 miles (8 kilometers) high. They formed *black blizzards* that swept all the way to the Atlantic coast.

def•i•ni•tion

The **Dust Bowl** was an area in the south central plains of the United States that was devastated by drought in the 1930s. **Black blizzards** is a term that was used to describe the dust storms that swept the plains during that time.

Wind erosion on this scale was something farmers in the area hadn't experienced before. Many pulled up stakes and moved to California. John Steinbeck catalogued some of the misery of this era in his novel *The Grapes of Wrath* in the late 1930s, when thousands of people migrated from the south central states to California, desperately searching for work. Others stayed to salvage the land. Federal and local soil conservation districts taught farmers how to rehabilitate the soil, and millions of acres were restored to grass. But farmers' memories were apparently short-term, because when rains returned in the 1940s, many plowed up the grasslands again. In the 1950s when another drought hit, Congress moved in and paid farmers to restore wheat land to grass and to instill soil conservation measures. These included alternating fields each year to restore moisture, using drought-resistant cover crops, and planting rows of trees to check the wind.

Changes Present

The twentieth century began hot, cooled a little in the middle (possibly from all our own pollution), and rose rapidly once again. Though scientists look at the last 150 years (the rise of industrialization) as the source of our global warming problems, the last 100 years saw a greater rise in temperature and CO_2 in our atmosphere than the previous 50, and the last 50 years saw the most accelerated rise of all. Scientists tell us that carbon dioxide and other manmade greenhouse gases in the atmosphere cause this warmth. So what has the modern warm period brought us so far?

The National Oceanic and Atmospheric Administration (NOAA) reports a reduced diurnal temperature range—the difference between daily high and low temperatures—over about 50 percent of the continents since the middle of the twentieth century.

Who's Got the Ice?

The past century has seen a dramatic retreat of glaciers worldwide. In an engraving made in the mid-1800s of the French village of Argentière, a glacier looms large over the town. In a photo in 1966, the river of ice has pulled well back into the mountains.

The use of nuclear submarines by scientists since the close of the cold war has allowed us to explore the Arctic Ocean in ways we never could before. The Science Ice Expeditions (SCICEX) programs spearheaded by the U.S. Navy and the National Science Foundation (NSF) have launched a number of undersea expeditions below the Arctic sea ice in recent years. They report that the Arctic Ocean's ice cover has thinned by an average of four feet—about 40 percent—since the 1960s.

If the Arctic continues to warm, the consequences could be grave. It is possible that the ocean's summer ice cover could completely melt in the coming decades. Although the possibility is remote, the consequences could be catastrophic. Climate models show that the absence of ice in the Arctic could completely change climate patterns in the northern hemisphere.

Changes in El Niño

And there are problems brewing down south. From 1976 to 1979, the behavior of El Niño in the South Pacific off the coast of South America changed. (For more information, see Chapter 11.) El Niños suddenly got more frequent and intense, and the cooler La Niña patterns dissipated. It is unclear whether this is due to global warming, but scientists believe that if global warming continues, El Niños will get even worse.

Changes in Antarctica

A team of NASA and university scientists studying satellite photos of Antarctica in January 2005 found evidence of extensive areas of snow melted in response to warm temperatures. It was the most significant melt observed using satellites during the last three decades. Combined, the affected regions encompassed an area as large as California.

The team was measuring snowfall accumulation and melt in Antarctica and Greenland from July 1999 through July 2005. They observed the melting in multiple regions, including areas where the melt had been considered unlikely. Water from melted snow can penetrate to the bottom of the ice sheet in some areas through cracks and channels in the ice. There, it can lubricate the junction between the ice sheet and the bedrock below—causing the ice to move toward the ocean faster and increasing sea level.

The Least You Need to Know

- Ice-age hunters only converted to agriculture after things got warmer.
- Although the weather has been nice for 10,000 years, small changes have made big differences.
- Changes in weather promoted the plague.
- The Little Ice Age (1400–1850) was a time of chill.
- In the twenty-first century, the ice is melting and El Niño is getting even worse.

The Dawn of the Industrial Age

In This Chapter

- ◆ Britain stokes the Industrial Revolution
- ◆ The steam engine hauls the loads
- ◆ Back in the USSR
- ◆ Riding in cars
- ◆ A Grand Canyon haze

The Industrial Revolution has brought about most of the manmade changes in our atmosphere. The term was first applied to the period in Britain from about 1750 to 1850. During those years, the steam engine and the rise of industrialization radically transformed commerce and society. The new machines could do the work of many hands, and economic output grew at phenomenal rates. From Britain, the revolution spread to Europe and North America and is underway right now in various nations in the southern hemisphere. The creation of metal, machines, and tools has fueled profound changes in society and in the atmosphere.

Early Changes in Britain

The Industrial Revolution had its roots in the scientific progress of the Renaissance (from the 1300s to the 1500s). Leonardo da Vinci sketched the precursors to the machines that would later be invented during the Industrial Revolution. Factories in Sweden were using water power as early as the 1720s. Gunsmiths in France were then developing their own factory systems. But it was in Britain that all the changes came together in the middle of the 1700s.

Feudalism began to break down after the Renaissance. The self-sufficient manorial village with its lords and peasants gradually gave way to commercial farming, where farmers took their products to markets to trade for cash. Market considerations replaced traditional practices, and things got more efficient.

In medieval times when repeated plantings exhausted the soils, farmers would let as much as one third of their lands lie fallow so that the soils could recover. However, in the 1700s British farmers began to use crop rotation, planting beans and grasses, which put nitrogen back into the soil. Property that previously lay empty part of the time could now be used continuously.

Crop rotation allowed farmers to use the land that had lain fallow for pasture, since the cattle could graze on the grasses. Thus, farmers didn't have to dedicate land strictly for pasture and could plow up old pastures, grow more crops, and raise more cattle. With more cattle, they had more meat to get through the winter, more leather, and more organic fertilizer.

Then they started putting up fences. English agriculture had previously run under an open-field system in which everyone in the village had equal rights to the open land. But Parliament let people stick up fences around their lands, so things got more competitive—and again, more efficient. The agricultural revolution set the stage for the industrial one to follow.

The rise of free-market trade happened in Britain much more rapidly than it did on the continent. The British nobility was hooked on trade. Britain had lots of rivers and seaports and an overseas empire with plenty of raw materials. Merchants, landowners, and ship captains were out hustling, investing, and making deals. Improvements in sanitation, health care, and agriculture helped the population expand. Europe had money to burn from the gold and silver it had stolen from the newly conquered Native Americans. British merchants looked at each other and said, "We've got to have ourselves a sale."

Textiles Lead the Way

To have a sale, however, the British had to have something to sell. Textiles must have
certainly come to mind. The British were famous for them. But making cloth and
clothing was such a slow process. There were a number of tedious steps that went
into producing a piece of cloth. The fibers had to be combed until they were parallel.
They needed to be spun or twisted to make yarn or thread. After that, you had to get
out your loom and hand-weave some cloth.

People had been trying for centuries to mechanize these processes. Lewis Paul
invented a spinning machine in 1738 that could mechanically shape fibers and spin
thread. Edmund Cartwright patented a crude power loom in 1785. The hand loom
operators burned down one of the first factories, but that didn't stop progress. By
1813, there were about 3,000 power looms in operation in Britain; 20 years later there
were 100,000.

The Steam Engine

It was the steam engine that really fueled the Industrial Revolution. By the early
1700s, there were already a number of crude steam-powered machines for pumping
water out of the mines. (Mines flood when they go deeper than the water table.) But
these early devices used massive amounts of coal and were only practical at mine pits
where coal was cheap. James Watt (1736–1819), a Scottish engineer, was called to fix
one of these big guys in the 1760s—and he started thinking about ways to make it
better. With the old engine, the cylinder had to be heated up and then cooled down
to bring about condensation. Watt devised a separate condenser attached to the cylinder. He made the engine reciprocating by letting steam into one end of the cylinder
and then into the other end, adapting this power to produce rotary motion. In other
words, it could turn a wheel.

More than that, by converting the steam power into rotary power, it could turn mills and operate machinery that had previously been powered by water wheels. Plus, the machine could be taken anywhere you wanted—and you didn't have to rely on the local creek to keep it going.

Coal and Iron

Before 1800, most machinery was made of wood. But wood splintered, warped, and cracked. If the Industrial Revolution was to get going, it needed some harder material for its equipment.

There was iron, but iron wasn't easy to create on a mass scale. Iron ore contains iron oxide, and you have to get the oxygen out to make metal. Up until then, iron was refined with charcoal fires. As the charcoal burned beneath the ore, it drew the oxygen off and left the pure metal. Charcoal was made from wood, and the forests of Britain were going down fast. So it was a big deal in the eighteenth century when the iron makers figured out how to use coal to make their iron.

With coal, you could get your iron molten but it still had some impurities—chiefly, carbon. This was wrought iron, and it could be poured into molds but was too brittle to make into hinges, swords, and hardware. To get the better grade of iron necessary for tools, you had to get the fire hotter—usually by blowing air on it to produce carbon-free metal. One of the first uses for the steam engine was to power these blowers to improve the production of iron. Steam engines were also used to power forging hammers and rolling mills. In the 1850s, industrialists learned how to convert iron metal into harder steel.

What's That Smell?

It wasn't long before the new factories began affecting the air that people breathed. But it was not the first time there had been pollution problems in Britain. As far back as the thirteenth century, Londoners had complained about the tanning, pottery, and lime production industries—all of which contributed their particular smells to the city air. Coal came to London soon after 1200 C.E.—and mounting sewage problems as well. In 1257, King Henry III's queen became the champion of the complainers—one of the first environmentalists—and from that time on, commissions were regularly established to consider air pollution.

However, as the forest dwindled and wood became more difficult to find for domestic fires, people turned more to coal. Nice people at first refused to enter rooms where coal had been burnt, but the snobbery died quickly as more people started using the black substance.

During the 1600s, rickets—a children's disease caused by a lack of sunlight—became increasingly common due to smoke-filled skies and factory labor for children. The 1700s and the 1800s didn't get any better. A survey conducted in the early 1900s found that half the children in the poorest districts had rickets. Other major European cities experienced similar problems when the Industrial Revolution was exported to their countries. But the increased capital that industrialization brought with it was apparently too enticing for people to object to a lack of sunlight. Rather than complain about pollution, they simply learned to live with it.

Global Warnings

You can chart the degradation of the air during the Industrial Revolution by looking at the art it created. Blue skies dominated the backgrounds of most of the early landscape painters in Britain and the continent. But as the decades progressed, backgrounds became hazier and the blue disappeared.

Steam Transportation

The Industrial Revolution really started moving when the steam engine was put on a railroad car. Rails laid on prepared beds had been used before, principally for the horse-drawn cars used for hauling ore and heavy machinery around the mines. It wasn't until the early 1800s that the first steam-powered locomotives were in use. It was a revolution in shipping. No longer did industry have to depend on the packhorse or the wagon. In 1807, Robert Fulton, an American who was educated in Britain, built the first steamboat and launched it on the Hudson River.

Britain wasn't afraid to borrow ingenuity from its former colony. There followed a heavy spurt of canal building in the British Isles. Canals linked many parts of the country and moved heavy freight at a fraction of the cost of hauling it overland. Roads kept improving as well, particularly after the first iron bridge was built in the late 1700s.

Industrialization Spreads

In the nineteenth century, Britain was still the center of the mechanized miracle. At first, it only exported products of the Industrial Revolution. It was the proverbial merchant with lots of goods to offer but little information on how it developed those products. However, it was a secret too lucrative to be kept. Soon, the revolution spread to Europe and North America. Then it spread to Japan, China, and India.

Pollution followed the spread. Visible smoke and the noxious fumes of sulfur dioxide (SO_2), a by-product of the burning of coal, followed wherever industrialization went. London fogs—a mixture of fog and pollution—were at times so thick that street lamps were turned on in the middle of the day. Charles Dickens called these fogs "London Particular." As pollution spread, so did the resistance to it. In the United States, laws regarding air quality were adopted as early as 1815 in Pittsburgh, Chicago, and Cincinnati.

By the nineteenth century, the citizens of London had banded together to protest the increasing problem. But the cries of those who had tasted the fruits of industrialization and wanted more drowned out those who were complaining about the pollution. The same story repeated itself in Europe and the United States. Industrialization pushed forward.

The Car

A French engineer built the first motorized vehicle in the late 1700s. It was a heavy steam-driven carriage with a big boiler that stuck out in front. It could travel a whopping 3 mph (5 kph). The Stanley brothers of Massachusetts, the best-known American manufacturers of steam-driven cars, produced their famous Stanley Steamers from 1897 until after World War I.

The development of the internal combustion engine fueled the rise of the automobile. Karl Benz, a German engineer, built the first vehicle of this type in 1885. French, British, and U.S. imitations followed. In the United States, the first horseless buggies were manufactured in the 1890s. Many of the early engines had only one cylinder and a chain-and-sprocket drive that pushed wooden carriage wheels. They were open (like the buggy), accommodated two passengers, and were steered with a lever.

Automobile production didn't take off right away. George Selden held a patent on the automobile with a group of manufacturers. Henry Ford, however, represented a group of independent manufacturers who refused to acknowledge the patent. Selden's group sued Ford and his cohorts. In 1911, the U.S. Circuit Court of Appeals ruled that the patent only covered the two-cycle engine. Ford's engine was a four-cycle. Ford then got the assembly line going, and soon the automobile was linking cities and suburbs all across North America. Trucks brought luxuries to places that hadn't known them before. Together, cars and trucks began to transform the landscape and society—as well as the air.

The Ruhr and World War

Industrialization heated up on the European continent, particularly around both World Wars. Much of the activity in Germany was centered in the Ruhr region, near the junction of the Ruhr and Rhine Rivers. In 1850, this had been an agricultural area—but then a rich seam of coal was discovered beneath it. By 1910, the region employed hundreds of thousands of miners, digging up millions of tons of high-sulfur coal. The coal sustained the giant steel and ironworks of Krupp and Thyssen—the muscles behind the German military-industrial complex.

At the start of the twentieth century, the Ruhr was the biggest industrial region in Europe. Because this industry was so vital to the German military machine, it escaped most regulation. Soon the air was polluted with smoke, soot, and sulfur dioxide.

Under Hitler, Ruhr factories stoked their furnaces even higher. Though the Nazis expressed affection for German blood, that affection apparently didn't extend to German lungs or German air. In 1944 and 1945, Allied bombers made special targets of the Ruhr—eventually leveling a large part of its factories. The haze, however, made Allied bombers far less accurate.

> **Global Warnings**
>
> Throughout much of the first half of the twentieth century, the Ruhr region—the industrial heartland of the German military machine—was so polluted that many came to believe the pollution was natural. In 1923, French and German troops occupied the Ruhr in response to Germany's failure to pay war reparations. Strikes followed, and the industry was shut down over the summer. The sky got bluer, trees grew faster, and the local harvest improved by half. There really was life without pollution. But then strikers settled, went back to work, and everything got hazy again. An official inquiry into the polluted air—inspired by the blue summer of 1923—concluded that the pollution was inevitable and warned that the people must adapt, rather than try to control industry.

Communism and Pollution

In the 1960s, air pollution became an important political issue in both Germany and in North America—but not in the USSR or any of its Eastern European satellites. High, billowing smokestacks were a symbol of pride for Marxist regimes—a promise of power and growth for its society. Iron, steel, and coal were the cornerstones of the economic plans of the Communist countries.

Air pollution was the norm. It was viewed as a symbol of progress rather than a plague of industry. Citizen pressure and pollution control, which were growing in the Ruhr toward the later part of the twentieth century, were absent behind the Iron Curtain. The state protected its polluters. Popular dissent wasn't allowed behind the Iron Curtain.

In the 1960s, when the German citizenry of the Ruhr Valley began to complain about pollution, the first response by industry was to put up colossal smokestacks that had the immediate result of reducing the emissions locally but broadcasting them over a much larger area. During the Communist rule of the USSR and Eastern Europe, these tall smokestacks were perceived by the Soviets as a symbol of power. It wasn't until the 1990s, after the fall of communism in Eastern Europe and Russia, that the cries against pollution were finally heard.

The Japanese Story

Japan followed the United States and Britain and from the 1870s on built heavy industrial centers. The nationalism of Japan and Germany in the early part of the twentieth century promoted industrialization as both countries expected their citizens to endure pollution for the good of the national cause. Both countries quickly became some of the most polluted in the world. However, some complaints were heard. Local farmers got miners to curtail operations during the rice harvest. In response to local complaints, the Hitachi copper mine in the 1910s built the tallest smokestack (490 feet, 155 meters) in the world.

The automobile took over in the second half of the twentieth century. Tokyo's car fleet quadrupled from 1960 to 1970. When the head of the American Environmental Protection Agency (EPA) went before Congress to push the Clean Air Act in 1970, he showed them photos of Tokyo traffic cops who had to wear masks to filter the air. It was an example of how bad things could get in the United States if Congress didn't enact tougher regulations.

But by 1990, Japan was not the same polluted nightmare it had once been. Part of the reduction in pollution came from the switch from coal to oil and other energy sources. The Japanese economy was then booming, and Japanese industry had enough capital to invest in pollution control. It also invented cars and machinery that were more fuel efficient. Japanese industry was able to continue its growth while decreasing its fuel consumption.

When the U.S. Clean Air Act was being enforced in the 1970s, Japan's powerful Ministry of International Trade and Industry came up with its own environmental restrictions. But whereas American automakers sued the EPA, Japanese automakers complained but then dove into the task of meeting both the new Japanese and American standards. The result was that by 1978, new Japanese cars emitted only about 10 percent of the pollution of the 1968 models. While the major American automakers dallied, Japanese car manufacturers surged into the American market in the 1970s.

Estimates vary as to the effect air pollution has had on the health of the world's population during the last century. In 1996, Harvard University's School of Public Health estimated that air pollution killed 568,000 people annually. In 1997, the World Health Organization (WHO) estimated that the annual world death toll was closer to 400,000. One estimate of the human toll of pollution during the twentieth century puts the total deaths at between 25 and 40 million—roughly equal to the combined casualties of World Wars I and II.

Acid Rain

Acid rain, derived from sulfur and nitrogen oxides, became an international issue with neighboring countries in the twentieth century. Unlike most pollutants, sulfur and nitrogen oxides are long-lasting in the air and are capable of traveling thousands of miles. The combustion of fossil fuels in the industrial centers of the twentieth century generated huge emissions of sulfur—the main ingredient in acid rain. Sulfur and nitrogen oxides released into the atmosphere by the burning of fossil fuels combines with water vapor to produce sulfuric acid and nitric acid. Acid rain comprised of these elements can drift out over larger areas and decimate forests, fish, and fresh waters.

def•i•ni•tion

Acid rain is a general term for many phenomena including acid rain, acid fog, acid sleet, and acid snow. It is produced when emissions of sulfur and nitrogen from the burning of fossil fuels combine with the moisture in the air to produce a mild solution of sulfuric acid and nitric acid that is destructive to plant life, aquatic life, monuments, and buildings.

One of the consequences of industrialization is the rise of the unhappy "downwind club," whereby nations that don't reap any of the benefits of their industrial neighbors nevertheless share in their pollution—principally delivered as acid rain. In the 1990s, industrial pollution from Korea and China rained down on Japan. Acid rain produced around the Great Lakes and in the Ohio River Valley fell on Quebec (as well as New York and New England). It was the same case in the 1960s, when British pollution spoiled the rivers and lakes of Scandinavia. In 1985 and 1994, most European countries signed protocols requiring sulfur emission reductions. Emissions dropped 15 percent from 1980 to 1995, driven mainly by the energy efficiency stimulated by petroleum price hikes in the 1970s.

Acid rain and the acidic brew that comes from smokestacks and tailpipes corrode limestone and marble and have thus taken a toll on our cultural monuments during the twentieth century. Acidic London fogs have eaten away at the stone facing of St. Paul's Cathedral. Athenian pollution has done more damage to the ancient marble on the Acropolis in the last 25 years than all the weathering of the previous 2,400 years. Michelangelo's David could no longer be trusted to Florence's acidic air and was replaced by a replica outdoors. And the Sphinx, which has survived centuries of desert sandblasting, showed its greatest degradation in the twentieth century as a result of Cairo's pollution. How many of these monuments will even be recognizable in the decades to come?

Smog

The experience of Los Angeles in the 1940s was a harbinger of pollution problems that would occur elsewhere in big cities across the nation. When the residents of Los Angeles complained about the smog, few people were aware at that time that smog—which derived its original meaning from a combination of smoke and fog in London—was indeed different there. Los Angeles industries did not use coal, yet the air was still hazy with pollution. What scientists discovered was that Los Angeles smog wasn't due to smoke and fog but instead to the action of sunlight on gases emitted chiefly from car exhausts.

Los Angeles got rid of its trolley system in the early part of the twentieth century and banked its hopes on a system of freeways that emphasized the automobile as the chief mode of transportation. As air pollution problems have increased, the city has tried to reinstall its trains as well as add subways and buses. But the expenses are far greater since the city tore up its rails (and in many cases sold the right-of-ways). Smog is also a problem in the Baltimore to Boston urban corridor—but at least on the East Coast, mass transit is well established.

Our National Parks

From the Great Smoky Mountains to the Grand Canyon, air pollution is seriously affecting our national parks. Clouds over the Smoky Mountains can be 1,000 times more acidic than natural rainfall. More than 30 species of plants in the park show injury from ozone. At the Grand Canyon in Arizona, pollution spoils the views on 90 percent of the days. Visitors to the south rim can only see half as far across the deep red canyon as when the park was dedicated in 1919.

Figuring out where the pollution all comes from is not an easy task. In the Great Smoky Mountains, pollution can originate from power plants in the Tennessee or Ohio River valleys—or it can drift in at high altitudes all the way from Atlanta or New York. A recent study in the Grand Canyon showed that its pollution came from industry in Arizona, cars in California, smokestacks in New Mexico, Nevada, and Utah, and unregulated pollution sources as far away as Mexico.

Et Tu, Global Warming?

And now we add global warming to the mix. In many cases, it originates from the same band of desperados who are the villains of ground-level pollution: ozone, carbon monoxide, particulates, and sulfur oxides. All have greenhouse effects. But the principle villain of the global warming problem is carbon dioxide. Unlike the other pollutants, it is colorless and odorless but is far longer-lasting.

Pollution can change the weather locally and also in areas that are downwind from pollution sources. Suspended particulates act as condensation nuclei upon which the moisture in the air gathers to form clouds, rain, snow, hail, and fog. In Pittsburgh, Pennsylvania, and Buffalo, New York, rainfall is generally lower on Sundays when industry and automobiles are running at lower volumes. In Toronto, Canada, the snow is heavier when there is more particulate matter in the sky.

However, as bad as our pollution problems are, they do not have the capacity to change the climate—as does global warming. Yet pollution comes from the same source as global warming: the same growth in industry, the same automation, and the same cars. It is another dangerous side effect of our ravenous consumption of fossil fuels.

The Least You Need to Know

- The Industrial Revolution occurred first in Britain from 1750 to 1850 and has since spread throughout the world.

- Both the Nazi and Communist regimes encouraged industry at the sacrifice of the air.

- Japan experienced a reversal of pollution problems in the latter half of the twentieth century.

- Pollution has no problem jumping international borders.

- Pollution and the manmade greenhouse gases responsible for global warming have the same source: the burning of fossil fuels.

Chapter 8

Trees and Carbon Dioxide

In This Chapter

♦ Forests covered the land

♦ Man removes the covers

♦ Rainforest wealth

♦ Rainforest devaluation

Humans have been hacking back the forest since they first decided to take up farming and ranching. There are various estimates of the history of forest destruction by humans given by environmental groups. Anywhere from 20 to 80 percent of the forests that existed in pre-agricultural times has been removed. Today, the biggest battles between humans and the trees are taking place in the tropical rainforests.

Rainforests are found around the world, mostly near the equator. About 80 to 90 acres of tropical rainforest are destroyed every minute. An area the size of New York State disappears each year. And as the trees are cut down, they release a tremendous amount of carbon dioxide into the atmosphere. The destruction of the forest and the tilling of the land produces about 25 percent of the greenhouse gasses released by man into the atmosphere.

The Rise of the Forest

Forests have been on the earth for millions of years, although their ancestors looked quite different. The fossil remains of the scale tree, actually a form of giant moss, are found in the rock strata of the Carboniferous period about 300 million years ago. These guys had weirdly shaped, skinny stems that could reach 100 feet (30 meters) in height. They occurred in the warm, moist lowlands that surrounded inland seas.

The first time that trees start to look anything like what we see today was about 100 million years later. Conifers and broadleaf trees began to form and diversify. Conifers are evergreen trees (they stay green all year) and have needle-shaped leaves. Broadleaf forests are generally deciduous trees (they shed their leaves each year) and have broad, flat leaves. Conifer and broadleaf trees represent the two major categories of trees in the world. Conifers are found in cooler, drier climates, and broadleaf forests are found in warmer, wetter areas of the world.

def•i•ni•tion

Conifers are trees with needle-shaped leaves that are evergreen (they stay green all year). Broadleaf trees are those with broad, flat leaves that are often deciduous—shedding their leaves each year.

About a third of the forested area of the world is coniferous—most of it occurring in the northern hemisphere. The area that encompasses the former Soviet Union contains 45 percent of the world's coniferous forest while North America contains about 35 percent. The continents of South America and Africa alone contain about three fourths of the world's broadleaf forest—much of it tropical rainforest.

Carbon Dioxide and the Forest

Carbon is one of the most important elements of the forest. Trees and other plants absorb carbon dioxide, convert it to food and energy, and release oxygen back into the atmosphere so that all of us oxygen-consuming creatures can keep sucking up air. We take in oxygen and exhale carbon dioxide. This cycle makes sure that the supply of carbon dioxide and oxygen remains fairly stable. But cutting down vast numbers of trees, whether in temperate or tropical forests, releases billions of tons of carbon dioxide into the atmosphere—either by decay or burning—that would otherwise remain locked up in the trees.

Forests act as carbon sinks, or storehouses of carbon, removing vast quantities of carbon dioxide from the atmosphere. One of the ideas proposed in the 1997 Kyoto Protocol was to allow industrialized nations to meet their carbon dioxide reduction goals by planting more trees. Instead of having to reduce fuel emissions by 10 percent, they could reduce emissions by 5 percent and plant enough trees to take up the difference.

Old-World Destruction

Humans began cutting down the forest when they evolved from nomadic hunters to settled farmers. As civilization developed, so did the need for timber. Man used wood for fuel, shelter, and other domestic uses. Wood was important for building mines and smelting metal. It was used early on to build dugout canoes and later for tall ships.

At the end of the first millennium, settlers had made deep inroads into European forests. Wood in the Middle Ages was an important part of the British military machine. Each foot soldier carried 24 arrows made from the wood of ash trees harvested specifically for that purpose. Bows were made of yew. By the 1600s, French officials were trying to regulate wood harvests so that enough good-quality wood was left over for shipbuilding. During the 1600s and the 1700s, woodsmen were already looking for ways to reproduce and sustain the forests.

Settling America

When the settlers reached America, their first thought was clearing the land for farms—not preserving the trees. Native Americans had a more sustainable relationship with the forest and the land. But the new settlers had farms to clear, grazing lands to develop, and industrial fires to stoke.

Clearing the Land

There were approximately 850 million acres of forest when European settlers first arrived in America, but quickly about 100 million acres were cut or burned to make way for farming. Between 1850 and 1900, while the United States was engaged in its own Industrial Revolution, an average of 8.6 thousand acres were cleared every day to make room for farms. By 1900, agriculture, construction, and fuel needs had reduced the original forest cover to fewer than 580 million acres.

Prior to 1900, North Americans were not that interested in preserving forests. Their main interests were expanding agriculture, roads, industry, transportation, and ranching. All of these activities contributed to the destruction of the forests—and the release of stored-up CO_2 that these forests held.

The Lumber Industry

As America expanded westward, forest products became big business. Frederick Weyerhaeuser, who was known as the "Timber King," was born in Germany and moved to the United States in the 1850s. He married and then moved to Illinois, where he worked in a sawmill. By 1857, he'd saved enough money to buy the business. Things went so well that he bought several more sawmills along the Mississippi.

In the 1890s, Weyerhaeuser moved to Minnesota where he struck up a friendship with James Hill, the railroad tycoon. Hill had lots of forest that he'd bought cheaply from the U.S. government—principally as right-of-ways for his railroads—and he didn't know what to do with it. Weyerhaeuser talked Hill into selling him three million acres. The purchase seemed to whet Weyerhaeuser's appetite for timberland—a habit that eventually led him to own more forest than any other American. Weyerhaeuser's practices have come under fire by environmentalists in the Pacific Northwest and more recently in the tropical rainforest, where the corporation has expanded its timber empire.

About the same time that Weyerhaeuser was gearing up his timber holdings, the U.S. government was getting involved in forestry practices to try to guarantee a sustainable yield of timber. Reforestation was first practiced in Germany in the 1300s. By the mid-1700s, it was well established in France, Switzerland, and Austria. German immigrants brought modern forestry practices to North America in the 1800s.

The timber industry has changed in recent years as large multinationals have taken over smaller timber companies in the Northwest and have sold off timber rapidly to pay debts or to finance other more profitable endeavors. Many timber companies now ship whole or split logs to Japan, Asia, and Mexico to be milled.

The Conservation Backlash

The idea of conserving our nation's forests got its first supporters toward the end of the 1800s—long before scientists realized the importance of forests as a sink for greenhouse gas emissions. In 1892, the same year that John Muir established the Sierra Club, Congress added a rider to the U.S. General Revision Act—authorizing the president to set up forest reserves. Gifford Pinchot, one of America's leading advocates of forest conservation, was declared head of the Division of Forestry. Pinchot was born into a wealthy Connecticut family, graduated from Yale University, and was encouraged by his father to get into forestry. No school in the United States or Canada taught forestry at that time, however, so Pinchot went off to France to see what he could learn.

In 1905, Teddy Roosevelt established the U.S. Forest Service and appointed Pinchot the chief forester. At that time, lumbermen regarded forest devastation as part of the job and the idea of rejuvenating the forest as a delusion. Pinchot emphasized long-term forest management and the wise use of timber resources. During Pinchot's five-year term as forest chief, the number of national forests more than doubled.

In the second half of the twentieth century, increasing attention to preserving old growth forests—forests that have never been logged—has created intense controversy between environmentalists and logging communities. Environmentalists argue that old growth trees are necessary to some species of animals—particularly the endangered spotted owl, which scientists believe need the more mixed, old growth forests to sustain themselves. Loggers argue that the real endangered species are the logging communities that rely on cutting old growth forest for income. National parks and wilderness areas protect much of the old growth forests in North America today.

def•i•ni•tion

Old growth forest is a type of forest that has not been previously logged and represents a more natural and diversified mix of tree and plant species.

The Rainforest

Today, the tropical rainforests of the world are under the gravest assault. Tropical forests are vanishing at the rate of 250 acres per minute. This means that an area roughly equal to two football fields (about four acres) disappears every second. An example is Mexico, which has lost half of its temperate and tropical forests since the 1960s.

Tropical rainforests are composed of three different types of forests: tropical dry forest, tropical moist forest, and true tropical rainforest. True tropical rainforests receive about 80 to 120 inches (203 to 305 centimeters) or more of rain per year. Tropical rainforests cover about 6 to 7 percent of the earth's surface and are located in a band reaching 10 degrees in both directions from the equator and covering about 3.4 million square miles.

Tropical rainforests have layers of vegetation that form separate habitats for different animals. At the top of the forest is the emergent layer, where a few giant trees thrust their huge, mushroom-shaped crowns above the dense canopy area. The canopy layer is where the crowns of the rainforest trees form a tight, continuous canopy 60 to 90 feet above the ground. The branches are often covered with moss, ferns, orchids, and other flowering plants. This is the layer where eagles, hawks, sloths, parrots, and monkeys hang out. The understory below receives only minimal sun and is the layer where young trees and leafy herbaceous plants that tolerate low light are found. The forest floor receives the least sunlight and has the fewest plants. Unlike temperate deciduous forests, only a thin layer of decaying matter is normally found here.

Wildlife Treasures

Tropical forests contain a larger diversity of plant and animal species than any other place on earth. Scientists estimate that anywhere from 50 to 90 percent of all animal species are found in tropical rainforests. Normally, animals are spread out and hard to spot. They rarely congregate. I got to see an exception to this rule at Santa Rosa Park in Costa Rica.

I arrived there in the dry season—a period during which the park's several waterholes become increasingly important to its wildlife. In the early morning and late evening, I walked a trail lined with black iguanas that led to one of these holes. Iguanas are giant lizards that can weigh up to 7 pounds (3 kilos).

One day as I approached the waterhole, a pungent odor filled the air. I looked to find a pack of peccaries (wild pigs) stoically drinking. Wild peccaries are said to be the most dangerous of Central American animals, but the peccaries at Santa Rosa were apparently nonaggressive. As I watched, a group of coatimundis—a South American cousin of the raccoon—threaded through the peccaries to share the water. Doves alighted on what little space was available while turkey vultures soared overhead. Then, as a chorus to this strange animal show, a troop of white-faced monkeys appeared, perching in the tree—waiting their turn for a drink.

If that watering hole hadn't been in a protected park, it might be farmland or a used and abandoned space by now—with the carbon locked into the forest released as CO_2 into the air. Tropical rainforest animals, many of which haven't yet been named by scientists, are disappearing as fast as the forest.

A Cornucopia of Pharmaceuticals

One of the little-known treasures of the tropical rainforest is its medicines. About 25 percent of all pharmaceuticals used by North Americans were originally derived from tropical plants. Quinine, used to treat malaria and pneumonia, was developed from the bark of a rainforest tree. Important ingredients in cortisone are produced from rainforest plants in Central America. Anti-cancer medicines, contraceptives, tumor inhibitors, anesthetics, and drugs to stimulate the heart and respiratory system have come from rainforest products.

The National Cancer Institute sends *ethnobotanists*—scientists who study the relationships between plants and human cultures—to rainforest countries to investigate the plants used by native healers to treat their patients. The institute is concerned that if the rainforest disappears, so may the healers and their medicinal secrets. Many of these cures have developed over thousands of years through practical application: try this, see whether it works; if it doesn't, try something else. Although it's not the way Western medicine likes to develop its pharmaceuticals, that doesn't mean that a number of valuable cures may not have developed from the process.

In fact, there are currently more than 200 pharmaceutical companies in the world that are studying rainforest plants used by native healers and shamans to treat their communities. Many researchers believe the rainforest contains a treasure of chemicals that could be used as new pharmaceuticals.

def•i•ni•tion

Ethnobotanists are scientists who study the cultural and biological relationships of native peoples and the plants in their environment.

Indigenous People

Indigenous people are among the rainforest's most precious riches. There are approximately 250 million indigenous people in 70 countries that control about 10 to 20 percent of the world's lands. Most of it is in the harshest regions: the Arctic tundra, the dry deserts, and the most virgin rainforests. About 50 million indigenous people live in the rainforests of the world. Contact with outsiders can have drastic consequences. Many die when outside explorers enter their lands, carrying diseases for which the natives have no immunity. More than 30 percent of the Yanomami people in Venezuela have died from measles and whooping cough since their first contact with outsiders in the early 1900s. Perhaps the greatest toll took place in the 1970s when the Brazilian government built roads into the Yanomami territories, which attracted gold miners and other settlers.

Scientists believe there may have been an estimated 10 million indigenous people living in the Amazon rainforest five centuries ago. Today, there are fewer than 200,000. The diseases the Portuguese and Spanish explorers brought with them preceded the advance of the Europeans into the Amazon, wiping out the indigenous people before the explorers arrived in number.

Many indigenous rainforest communities have a sustainable relationship with the forest that has lasted for thousands of years. According to the Rainforest Foundation of Great Britain, indigenous people practice shifting agriculture, hunting, gathering, and fishing. They have an intimate knowledge of plants that they use to provide them with food, medicine, pesticides, poisons, building materials, clothes, string, and implements.

Recent satellite evidence shows that indigenous territories in the Amazon have a better record of preserving the forest and the land than even Brazil's national parks.

In many cases, this is due to the fact that indigenous people tend to prosecute trespassers—whereas parks don't have enough money or manpower to do the job.

Cutting It Down

Much of the fertile land in the Amazon Basin and the Philippines has been turned into large plantations to cultivate rice, sugar cane, bananas, pineapple, rubber, and coffee. The major timber companies have developed tree plantations in rainforests. This often changes the natural diversified forest into single-species stands that are raised in an intense agricultural setting, which includes chemical weeding, pesticides, and mechanized harvesting. These processes destroy soil fertility and groundwater reserves, and increase erosion.

In the late 1960s, the Brazilian government began a program to encourage development in the Amazon that included building roads. They offered settlers 240-acre plots with guaranteed credits for planting crops, a six-month household subsidy, and food subsidies as insurance against disasters. To encourage commercial investment, they offered tax incentives to businesses, subsidized loans, and reduced import duties on agricultural equipment.

Global Warnings

The $40 billion Avanca Brazil (Advance Brazil) project announced by the Brazilian government in 2001 holds special dangers for the Amazon. The program promotes industrialization, highway construction, hydroelectric projects, and more. Researchers reported in *Science* magazine that the enormous project could degrade 28 to 42 percent of the Amazon rainforest by 2020.

Martyrs for the Rainforest

Chico Mendes was murdered by cattle ranchers in 1988 for trying to protect the Amazon rainforest in eastern Brazil. Born in 1944, Mendes grew up in the rainforest. His family worked on a rubber plantation and grew their own food or harvested it from the nearby forest. By the time he was in his early 20s, Mendes was organizing rubber tappers for better working conditions, fair prices from company stores, and the building of schools. When the government started building roads and cutting down the forest upon which Mendes depended, he joined with other grassroots leaders and several activist Catholic priests to organize the workers.

By the 1980s, the movement had grown so strong that a conference held in Brazil in 1984 attracted more than 4,000 delegates from unions and other organizations for workers. At the conference, Mendes promoted the idea of setting aside large sections of the rainforest as reserves. Mendes received the United Nations' Global 500 Award for environmental conservation in 1987. But his enemies were enraged at his successful efforts to stop rainforest destruction. His death brought worldwide attention to the conservation of the rainforest.

In 2004, Dorothy Stang—a 74-year-old nun from Dayton, Ohio—was assassinated on the street in broad daylight for her opposition to rainforest logging in another state in the Amazon. According to a witness in her car, gunmen had stopped the car and were pointing their weapons at her when she took out her Bible. She claimed that it was her only weapon and started to read passages from it. The gunmen fired six bullets at her at close range. Protecting the rainforest has become a dangerous cause to champion.

Oil and Roads

In the last several decades, oil companies have explored various regions of the tropical rainforest in South America with the hopes of finding new reserves to feed the voracious petroleum appetites of North America and Europe. When oil companies find petroleum, not only do the derricks and pipelines pollute the soil and the waterways—but oil companies also build roads into the forest, which are later used by poor farmers as access to clear more land.

Ten years ago, I flew by helicopter into Vilcabamba, a rainforest in a mountainous region of Peru and a part of the Andes, between its peak and the Amazon Basin. The purpose of the expedition was to explore the region and survey its species. It was sponsored by Conservation International, a Washington, D.C.–based environmental group.

The Shell, Mobil, and Chevron corporations had won concessions from the Peruvian government to explore the valleys just east of Vilcabamba and to build a pipeline to Lima over the Andes. Under this eminent threat, Conservation International decided to send in its Rapid Assessment Program (RAP) team, a group of American and Peruvian biological superstars (biologists from the Smithsonian Institution, the Field Museum of Natural History in Chicago, and the Museo de Historia Natural in Lima, Peru). The goal was to survey the wildlife of Vilcabamba in four weeks. Scientists believe the tropical Andes are among the most biologically diverse of all the remaining forests in the world. Vilcabamba, an uninhabited wilderness in the midst of this range—roughly the size of New Hampshire—has enormous conservation potential.

It rained every day during my stay in the Vilcabamba rainforest. In the morning, there were brief periods of sunlight during which the biologists would hang wet clothing on every available tree, tent, or rope in desperate hope of drying them out. They'd pull them in quickly when the clouds started to build. The forest was so dense you couldn't see that many animals until the biologists put out nets that would quickly fill with tropical birds by day and bats by night (which biologists would catalog and release). They also spotted white-faced capuchin monkeys, spider monkeys, and night monkeys. And they found signs of puma, dwarf deer, and bear.

Climatoids

The tropical Andes is what scientists refer to as a hotspot—a term coined to identify areas with large numbers of species found nowhere else on Earth that are concentrated in a relatively small space. It's thought that if we can at least save these areas, we will be able to preserve a maximum number of species given the limited amount of conservation dollars.

The biologists discovered a number of plants, animals, frogs, and butterflies that had never been recorded before. Tom Schulenberg, the RAP team leader, was the local bird expert. In the weeks that we were there, he worked the edge of the bog in the middle of the forest, training both binoculars and microphones into the trees. According to Schulenberg, you can hear approximately three to four times more birds than you can see in the rainforest—and he can recognize the calls of different birds. He believes that Vilcabamba has at least 650 species of birds, roughly equivalent to the total number of species that breed in the United States and Canada combined.

Mobil, Shell, and Chevron are making major efforts to promote their new ecological awareness, and scientists are working with them. But they are cautious because Mobil made such a show in Tambopata, a mecca for bird watchers and a proposed national park in southeast Peru. At the last minute, Mobil walked off with a concession to explore the heart of the park, and conservationists got only a third of the space originally proposed.

More recently, I traveled with Wake Forest University biologist Miles Silman into the cloud forest in Manu National Park (I discussed this in Chapter 1). Silman believes that global warming is pushing the rainforest up the mountain faster than it can adapt.

Slash and Burn

Many rainforest farmers practice slash-and-burn agriculture. When they cut and burn the forest, it releases nitrogen and other fertilizers held by the trees back into the soil.

Farms cleared this way are usually productive for a few years, but then the soils are exhausted—and the farmers abandon their farms and clear more tracts. If allowed to lie long enough, cleared forest like this will rejuvenate. But the problem is that competition for arable land is so intense, the land is not allowed to rejuvenate—and farmers cut down more and more forest.

Brazilian scientists monitoring the rainforest by satellite estimate there are more than 8,000 fires in a single day in the Amazon rainforest. For the last two decades, the annual emissions of carbon dioxide from the burning of forests in Brazil has equaled some of the world's worst polluters. Brazil is the world's eighth-largest emitter of CO_2, and most emissions come not from industry and cars but from loggers, ranchers, and farmers burning the forest. Brazil has little industry. Without the burning, Brazil would not even make the list of top 20 polluters.

A Double Burger and a Side of Rainforest

Cattle ranching takes its share of the rainforest in Central and South America. Ranchers clear the rainforest and plant pasture for cattle. However, as with rainforest farming, the soils are soon exhausted and more forest must be cleared. During the 1970s and 1980s, much of the beef raised on these rainforest cattle ranches was shipped to Canada and the United States to be used as hamburger for fast-food restaurants. In the 1990s, both McDonald's and Burger King agreed not to buy beef from lands that were previously rainforest.

According to a 1997 report from the Food and Agriculture Organization of the United Nations, the transformation from tropical forest to crops and pastureland brings about substantial soil erosion and loss of soil fertility. Pastures cleared from rainforest can only be sustained for about 10 years. More than 50 percent of the pasture areas in the Amazon rainforest have now been abandoned in a degraded state.

Black Earth

Archeologists believe that the indigenous people who once occupied the forests of the Amazon, particularly along its major rivers, used a form of agriculture that differed from slash and burn. Trees were more sparingly cut and the wood that was felled burned more slowly, turning it into charcoal. The charcoal then was mixed into the soil, and this retained nutrients far better than ash. And it also retained more carbon. If current settlers would use this method, rainforest destruction could be dramatically reduced—and so could the release of greenhouse gases. (I'll tell you more about this in Chapter 23.)

The Least You Need to Know

- ◆ Forests once covered most of the earth but have been reduced by as much as 50 percent.

- ◆ Industrialization and the clearing of land for agriculture depleted the temperate forests of North America and Eurasia.

- ◆ The rainforests of the world make up a diverse storehouse of plants and animals.

- ◆ The rainforests are disappearing at the rate of two football fields (about four acres) every second.

Part 3

A Meteorological Primer

If you want to talk about global warming, you have to understand the weather—at least a little. In this part, I'm going to bring you up to speed. I'll tell you how winds are formed, how rain is made, and how hurricanes are born. I'll tell you the difference between weather and climate. (Hint: Weather is what we expect over the next 10 days. Climate is what we expect over the next 10 years.)

I'll talk about El Niño. What can a few added degrees of warmth do to the weather? El Niño is a good example. I'll take a look at some of the other climate variables that could be creating the warming we're experiencing. What if it's just sunspots?

Last, I'll take a look at some of the variables—like deep ocean current, the permafrost, and melting polar ice—that could throw the whole atmospheric system out of whack and ratchet up the global warming equation.

"Does this mean fewer snow days?"

Chapter 9

Weather 101

In This Chapter

- ◆ Where we get clouds
- ◆ The lows and highs of atmospheric pressure
- ◆ From drizzles to hurricanes
- ◆ Global air circulation

When we talk about global warming and its effects, we are essentially talking about how increased temperatures worldwide might change the weather and the climate. *Weather* is the condition produced by the temperature, humidity, wind, and precipitation at a given moment. *Climate* represents the weather averaged over a number of years. When we say the climate is getting warmer, we are not talking about the weather on a given day or in a given place—we're talking about the weather averaged over the entire world and for a number of years (decades, even). Because you were shivering last night in Alberta or sweating midday in Death Valley doesn't prove or disprove global warming. When we refer to global warming, we are talking about averages or climate.

Climate does affect the weather, though. Changes in globally averaged temperatures will have effects on rain, snow, sleet, and hail as well as tornadoes and hurricanes. It will also affect which areas get wetter and which ones get drier. But because warmer water produces more rain clouds and warmer air holds more moisture, the resulting storms may be more torrential. In 2005, while the United States was suffering from an onslaught of hurricanes, Europe was suffering from disastrous floods. Parts of Africa, on the other hand, have been suffering from persistent droughts. Over the last 40 years in the tumultuous Darfur region, Lake Chad—once the sixth-largest lake in the world—has almost disappeared. To understand more about the bigger climate picture, let's take a look at those things that affect the weather right now.

Temperature and Humidity

Good old water, H_2O, exists in three different states: as a solid, a liquid, and vapor. The temperature of the air determines in which state you'll find it. When the temperature falls below 32°F (0°C), water molecules stop moving freely, bind together, get hard, and turn into ice, snow, sleet, and hail. When the temperature gets hot, water turns to vapor. The molecules start moving freely, drift apart, and become clouds, fog, or humidity.

If the temperature is warm, the air has a greater capacity to hold water as vapor. During the winter months, the ability of the atmosphere to retain water vapor is almost four times higher in the tropics than it is in the midlatitudes because of the differences in temperatures.

When the capacity of air to hold water is reached, the excess water will fall out in some form of precipitation. Two things can create precipitation: either more water vapor is added to the air or the temperature drops, which decreases its capacity to hold that vapor.

One important term you might hear when listening to a weather report is *relative humidity.* This is the ratio between the actual amount of water vapor in the air and its capacity to hold that vapor. The relative humidity of the air increases if more vapor collects in the atmosphere or if the temperature is lowered. When relative humidity reaches 100 percent, it rains, snows, hails, or does something else to spoil your picnic.

def•i•ni•tion

Weather is the current condition produced by temperature, humidity, wind, and precipitation. **Climate** refers to an average of those conditions over a number of years. **Relative humidity** is a measure of the air's capacity to hold moisture versus the actual moisture present. To get rain, you must have 100 percent relative humidity.

Clouds

But getting from a state of 100 percent relative humidity to rain or some other type of precipitation isn't that easy. It has to first form into clouds, and there are a lot of complex processes involved. (Why else would we have the Weather Channel?) In fact, the water vapor in the air has a difficult time coalescing into cloud droplets or raindrops. The surface tension around the tiny vapor droplets is fairly strong. By themselves, water vapor molecules are not attracted to each other enough to form cloud droplets or rain droplets.

That's where those particulates in the atmosphere (which I mentioned in Chapter 2) come into play. Their role in the water vapor-to-raindrop problem is to act as nuclei around which cloud droplets can form. These particles come from ocean spray, forest fires, volcanoes, and smokestacks. That's why it rains more around polluted factories. Even when the atmosphere looks crystal clear, there are still plenty of these microscopic particles in the air. In a volume no larger than a finger, there could be more than 100,000 of these particles.

There are different kinds of clouds. The four basic types are *cirrus, stratus, cumulus,* and *nimbus.* Cirrus clouds are high, thin, and wispy. The high altitude and cold creates ice crystals, which cause the wispy effect. Stratus clouds are flat and are found closer to the ground. They like to drizzle. Cumulus clouds are scattered and puffy. Nimbus clouds are any clouds that deliver rain. (Cumulonimbus clouds are cumulus clouds that contain rain.)

The elevation at which these clouds arrive is important. High clouds—more than 20,000 feet—may foretell the arrival of a storm. Middle clouds—from 6,500 feet to 20,000 feet—are thicker. When high clouds give way to middle clouds, rain is approaching. Low clouds form at elevations below 6,500 feet and come filled with rain or snow. But low clouds generally deliver their precipitation steadily, not as thundershowers. Clouds that rise vertically, frequently into thunderheads, occur throughout the various levels. They are associated with strong, upward currents. They have a puffy—even towerlike—structure. Precipitation is heavy but not that long-lasting. Generally, the bigger they grow the more violent the weather becomes.

def•i•ni•tion

> Cirrus clouds are high and wispy. **Stratus clouds** are flat and low to the ground. **Cumulus clouds** are either puffy and unthreatening or gathering, rising, and thunderous. **Nimbus clouds** are rain-bearing clouds.

Precipitation Wears Many Hats

Precipitation comes in a variety of forms, as shown in the following list. Temperature is important as well as the upward movement of the air.

- **Rain** Rain is formed when cloud droplets coalesce. Over the ocean, the salt in the air helps the cloud droplets bind together—but over land, ice and microscopic particles (particulates) do the job.

- **Snow** Snow develops when ice crystals link to each other. Although snow is beautiful, it plays havoc with urban life. Lines form at the grocery store when the weather forecaster predicts snow. Snow comes in an infinite variety of crystalline shapes.

- **Sleet** Sleet is frozen rain that occurs in winter when warm air rises over a cold layer. Rain falls from the warmer upper layer through the colder lower layer, which is not quite cold enough to cause snow but cold enough to convert the rain to frozen pellets.

- **Hail** Hail is more of a warm-weather phenomenon. Hail forms when ice falls through clouds with vertical development but that have updrafts that keep sending the hail stones aloft, creating new layers of ice until they get so heavy they fall from the sky.

- **Freezing rain** Freezing rain turns to ice upon contact with cold surfaces. These rains are usually caused by ground-level freezing winds. This creates a dangerous glaze of ice that coats roads and walkways.

Why It Rains

Getting from water vapor to rain is complicated. First, the air has to be at approximately 100 percent relative humidity. We get there by adding more moisture or reducing the temperature. But it turns out that just bringing in more water won't get the microscopic droplets to bind together. We need a drop in temperature.

The surest way to lower the temperature is upward movement of the air. Remember from Chapter 2 that it gets colder as we go upward in the lower atmosphere. As it gets colder, the relative humidity gets higher. At 100 percent relative humidity, clouds form. If it gets high enough, ice forms—even during the summer and even in the tropics. Rising air is your best bet for rain or any of its relatives.

The upward velocity of the air affects how the moisture will develop. If the air is rising slowly, it may only form into clouds or light drizzle. If it's going up fast, it might fall out as heavy rain or a torrential downpour. Dense, rapidly building clouds are the stuff of thunderstorms and tornadoes.

Rising air produces clouds and rain, but falling air produces deserts. Such are the conditions on the West Coast when winds off the Pacific Ocean hit the Sierra Nevada Mountains. Rising air creates precipitation, which is why the western half of the Sierras is covered with trees and lakes. East of the Sierras is Death Valley. The same situation exists in Washington State and Oregon, where rain rising over the Cascades creates a densely forested area along the coast but a desert inland.

Under Pressure

Rising air is also where the low-pressure areas are, which makes a nice segue into the next topic: pressure. If you are feeling a little pressed lately, it might be the atmosphere. Actually, 10 to 20 tons of atmosphere rest on the average adult. The reason why we don't all collapse is that our bodies exert an equal amount of pressure from within. (Bet you didn't know you were that strong.) We actually go through life with this load on our backs and don't even notice it. Pressure is important in weather, though.

We measure pressure with a barometer. The barometer measures pressure in inches (millibars) of mercury. Normal sea-level pressure is 29.92 inches (1013.2 millibars). At ground level, atmospheric pressure fluctuates only a few percent—generally between 30.50 and 29.50 inches (1032.8 and 998.9 millibars). Because the air rises when it rains, the pressure will be lower. But severe storms will seldom have a pressure lower than 29 inches (982.0 millibars). The lowest pressure reading ever recorded for a hurricane (typhoon Tip, 1979) was 25.7 inches (870 millibars). Higher elevations have lower pressures because there is less air bearing down from above. Meteorologists speculate that the lowest pressure would be within a tornado, but nobody has figured out how to get a barometer to survive a tornado.

def•i•ni•tion

A **barometer** is a gauge used to measure the pressure of the atmosphere.

Low Pressure

Low pressure is where the action is … the storm action, that is. In low-pressure areas, the air is rising—which creates conditions for rain and its related wet friends.

Converging surface winds can cause low pressure. The winds come together, and there's no place else to go but up. Upward motion creates lower pressures because the upward-moving air is pushing against the weight of the atmosphere. Diverging currents in the upper atmosphere can also cause lower pressure, because air moving out from a given point in the upper atmosphere will cause air from the lower atmosphere to be sucked upward to fill the vacuum. Low pressure, therefore, is all about rising air currents.

Climatoids

The highest pressure ever recorded was 32.005 inches (1038.8 millibars) at Agata, Siberia, on December 31, 1968. It was 50 degrees below zero. The lowest sea-level pressure ever recorded was 25.69 inches (870 millibars) on October 12, 1979, about 100 miles east of the Philippines within Typhoon Tip.

High Pressure

High pressure is where the picnic and beach weather is. In high-pressure areas, the air is sinking and getting warmer. Downward motion creates high pressure because downward-moving air is adding to the weight of the atmosphere. High pressure is all about sinking air. And in a globally warming world, there will be more of it.

Storm Fronts

Storm fronts occur where air masses meet. Air masses are enormous bodies of air that have pretty much uniform moisture and temperature. A single air mass can cover one half of the United States. Air masses are given names based on the region in which they were born. Thus, air masses can be arctic, polar, tropical, or equatorial, depending on the direction from which they come—and either maritime or continental, depending on whether they were born out over the ocean (maritime) or the land (continental). Thus, a maritime polar air mass might form off the east or west coast of Canada, and a maritime tropical air mass might form off Mexico. A continental polar air mass might form in the middle of Canada; a continental tropical in the middle of Mexico.

Arctic and polar air masses are cold; tropical and equatorial air masses are warm. Maritime air masses are moist, and continental air masses are dry.

When these masses meet, you get a front. And fronts are trouble. They often exist where the storms occur. You can have cold fronts or warm fronts. A cold front is where colder, drier air overtakes warmer, moister air. Cold fronts create the most violent conditions. The cold, heavy air lifts the warm, lighter air rapidly. As the warmer, wetter air rises, tall clouds form and storms happen. Again, there is vertical motion of the air, which is the key to storm action.

Warm fronts occur when a warm air mass moves up on a cold one. Warm air masses don't move as rapidly. They slide up over the cold air masses more gently, at a more gradual angle. The clouds will be lower and the precipitation less intense. Sometimes all you get is drizzle or fog.

Doing the Cyclone

Frontal cyclones are those big, circular cloud formations you see from the weather satellites that everybody is always getting upset about. Wind veers to the right in the northern hemisphere and to the left in the southern hemisphere, creating the spin.

It's called the *Coriolis effect*. It's one of the forces behind storms. Those big, circular weather systems generally occur at the junction of two fronts. To get really big storms, you have to have diverging winds in the upper atmosphere that act as an exhaust fan, drawing the air up from below. Take a low-pressure zone, put some spin on it, add an exhaust fan, and you've got the makings of a whopper.

def•i•ni•tion

Frontal cyclones are large, spinning weather systems that occur along turbulent fronts. The **Coriolis effect** causes winds to veer to the right in the northern hemisphere and to the left in the southern hemisphere.

Tornadoes

More tornadoes occur in the United States than in any other country in the world. That's because the United States is caught between the cooler air masses of Canada and the warm, tropical ones farther south. Spring is the best time for tornadoes. That's when cold air masses still linger over the continent as warmer tropical air is moving up fast through Texas and the lower plains.

The lighter, warmer air rushes up over the lingering cold mass. A funnel can develop like a rolling log at the junction where the masses brush past each other. A strong downdraft can upend that spinning log, and a tornado can touch down.

Tornadoes can tear down a brick wall. A severe tornado can pick up a car and toss it in the air. Tornadoes destroy instruments that measure the wind, but researchers have gotten around that by using Doppler radar to estimate speeds—and tornado winds can exceed 300 mph (480 kph).

Most tornadoes in the United States occur in the region known as "Tornado Alley" that extends through the Central Plains states and the Midwest. There are 600 to 700 tornadoes reported each year in the United States. The most costly tornado incident was a cluster of tornadoes that ran through Iowa, Illinois, Wisconsin, Indiana, Michigan, and Ohio on April 11, 1965. It killed 271 people and caused $300 million in property damage. The most deadly tornado was a single one that ran through Missouri, Illinois, and Indiana on March 18, 1925. It killed 689 people. One of the effects of global warming could be an increased number of tornadoes.

The all-time record for the number of tornadoes in the United States was broken in 2004. The total was 1,717 tornado reports. It topped 1998 with the previous record of 1,424 tornadoes by almost 300. The year 2003 had a total of 1,368. The National Weather Service's historical records of tornado numbers date back to 1950.

In 2006, over a four-day period from March 9 to March 13, there were 105 confirmed tornadoes—making it one of the largest outbreaks in recent history. At least 13 deaths were blamed on the severe weather … 11 of them by tornadoes. The total damage by severe weather was in excess of $1 billion.

Hurricanes

Hurricanes are the mean guys—the worst storms we've got. Hurricane Katrina was the costliest and one of the deadliest hurricanes in the history of the United States. It was the sixth-strongest Atlantic hurricane ever recorded and the third-strongest hurricane on record that made landfall in the United States. Katrina formed on August 24, 2005, and devastated the central Gulf Coast of the United States. The worst loss of life and the most severe property damage occurred in New Orleans. That famous city, the only city in the United States that is below sea level, flooded as the levee system failed catastrophically. Katrina was the second Category 5 hurricane of the 2005 Atlantic season.

Hurricanes are called typhoons in the Western Pacific and cyclones in the Indian Ocean, but they're all the same animal. In 2004, 10 typhoons hit Japan (the all-time record). The previous record was seven. Until 2004, many scientists believed it was impossible to have hurricanes in the South Atlantic. But that year, the first hurricane ever hit Brazil.

Hurricanes are tropical cyclones whose winds exceed 73 mph. Unlike frontal cyclones, hurricanes don't occur along fronts of contrasting air masses. There aren't any big contrasts in the tropics; temperatures are all about the same. Hurricanes don't run on contrasting temperatures—they run on heat.

Most hurricanes that strike the eastern seaboard of North America originate off the northwest corner of Africa and move west over the Atlantic. They start with a cluster of thunderstorms converging. Remember, surface convergence creates low pressure because the winds are sent upward. Tropical heat assists the rise. Hurricanes form over oceans where the water temperature is at least 79°F.

Global Warnings

Hurricanes work their destruction by violent winds (Hurricane Andrew [1992], $40 billion in damage), torrential rains (Hurricane Mitch [1998], 11,000 dead), and enormous storm surges of water (Galveston Hurricane [1900], 6,000 dead). The total damages for Hurricane Katrina are estimated at $81.2, and its 30-foot-high storm surge killed 1,836 people.

As with frontal cyclones, they need that exhaust fan—the high-altitude divergence of air—to really get things going. Hurricanes are some of the few storm systems that might actually interact with the stratosphere. Hurricanes can cover an area with a diameter of 300 to 700 miles. A hurricane has spiral bands that radiate out from the center, and most of the rain occurs within those bands. The winds gradually increase and reach their greatest intensity in the eye wall, the area surrounding the eye of the hurricane. Some scientists believe that small tornadoes form in that eye wall and do the damage. The eye of the hurricane itself is relatively calm.

Sun Matters

Weather isn't all about water and wind: the sun also matters. The sun's heat collects in differing amounts around the surface of the globe, and this is the dominant cause of weather, modified by geography, terrain, and other factors. The tropical latitudes get lots of heat while the polar regions get little. The atmosphere and the ocean then try to even out the heat. Some of the equator's heat moves toward the poles, and some of the pole's cold moves toward the equator—trying to achieve a balance. Without these motions, the tropics would be even hotter and the poles even colder. As the air moves back and forth, storms form, sea breezes begin to blow, rain falls, and hurricanes develop.

Global Circulation

Although winds can change on a daily basis, there are some consistencies in world wind patterns. Working from the equator toward the poles, there are three bands of wind directions in each hemisphere. From the equator to about 30° latitude (about a third of the way to the poles), the winds come from the east and are called the trade winds. In the middle latitudes from about 30° to 60°, they come from the west and are called the prevailing westerlies. And from the poles, they come from the east and are called the polar easterlies. The Coriolis effect causes them to arrive at slight angles. Check out the following figure to get an idea of which way the winds blow. This model is a bit simplified. Winds don't always blow so uniformly. However, this gives us somewhere to start.

How the atmosphere circulates.

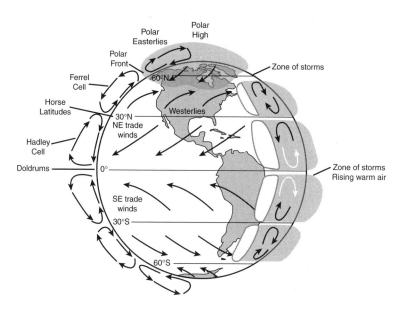

What regulates these winds are three distinctive cells or circular wind patterns in both the northern and southern hemispheres: one near the equator, one at the poles, and one in between. The trade winds from both hemispheres converge on the surface at the equator, which sends the air aloft—reducing the pressure. This is why it rains so much around the equator. When the air reaches the upper altitudes, some of it heads north and some heads south in an effort to even out the heat difference between the equator and the poles.

At about 30° latitude, the high-altitude winds sink to the surface. Sinking winds create high-pressure zones with lots of sun, which is why 30° latitude is where the world's great deserts are. Some of that sinking air heads back toward the equator, which keeps this cell or circular wind pattern going (see the previous figure).

Some of the high winds descend at the poles. From there, they head back toward the equator and converge with the prevailing westerlies at 60°. When winds converge, they rise. Thus, you get another area of frequent storms at 60°.

In general, there are a number of prevailing wind conditions on the surface and at higher altitudes in both hemispheres that greatly affect the weather. Where winds converge on the surface at the equator, the air moves upward and you will have lots of storms and rain. At 30° (one third of the way toward the poles), winds diverge at the surface, high winds move downward, and you get deserts and dry weather. At 60° latitude (two thirds of the way to the poles from the equator), you get converging winds and another uplifting wind with resultant big storm activity.

Climatoids

At 30° latitude, the surface winds diverge and air from upper altitudes must descend to fill the void. The descending air creates a high-pressure zone that's filled with sun, dry air, and not much wind. Over the oceans, this is called the horse latitudes. This is because at this latitude, sailing ships of old would encounter calm and the ships wouldn't budge. Sailors then threw everything overboard to lighten the load—including the horses.

Although surface winds vary at different latitudes, above 15,000 feet (4,560 meters) the wind is generally from the west. This can change from season to season—particularly in the tropics, where the flow is weak and winds can shift to the east during the summer. In general, though, these high westerly winds push weather systems along from west to east in both hemispheres.

So we have this complex system where air moves in complex patterns, stirring up trouble in the process. Now, what are a few extra globally warm degrees going to do? Well, a little extra heat is all it takes to get an El Niño going. We'll look at what that can do in the next chapter.

The Least You Need to Know

◆ Moisture and humidity go through changes to develop into rain.

◆ High-pressure zones are sunny and calm; low-pressure zones are cloudy and stormy.

◆ Rising air creates low-pressure zones; sinking air creates high-pressure zones.

◆ Where air masses with different temperatures and moisture content collide, look out for storms.

◆ The air over the globe moves in complex patterns created by the movement of air from the equator toward the poles and complicated by the rotation of the earth.

Chapter 10

El Niño

In This Chapter

◆ What El Niño can teach us about global warming

◆ Mudslides in California

◆ Ice storms in New England

◆ El Niño ups its ante

El Niño is a warm body of water that appears every two to seven years around Christmastime in the Pacific Ocean off the coasts of Peru and Ecuador. It usually disappears by the end of March but can persist up to 18 months. The increase in temperature isn't much: 2 to 3.5°F (1 to 2°C) in most years, 7 to 11°F (4 to 6°C) in strong El Niños. What's interesting is how close this is to the IPCC's prediction of global warming, which estimates from 2.0 to 11.5°F (1.1 to 6.4°C) over the next century.

Kevin Trenberth, a climatologist at the National Center for Atmospheric Research in Boulder, Colorado, thinks that El Niños may function as a pressure valve for the excess buildup of heat in the tropics. In a globally warmer world, he thinks we'll see more El Niños. There has, in fact, been an increase in the frequency of El Niños since 1976. But even if global warming is not the cause, El Niños provide us a neat working model of how a little additional heat may affect global weather.

What's Another Five or Six Degrees?

Gary Lemonoff will never forget the night during the El Niño storms of March 1998, when the rain-saturated earth above his home near Laguna Beach, California, gave way and tons of earth, rock, and trees slammed into his house. In the blackness of the night, he could hear his neighbors screaming. But the real horror awaited him in the early morning light when he looked out his window to see an arm and a leg protruding from the mud and rubble. Two people died in that slide.

But this was just a small portion of the devastation brought by the 1997–1998 El Niño, one of the most powerful climate events of the century. The United Nations estimates that massive storms, fires, floods, frosts, and droughts killed 2,100 people and caused in excess of $33 billion in property damage.

El Niño has been with us for a long time, but its potential to disrupt the weather worldwide was not systematically studied until the International Geophysical Year of 1957–1958. From this, we've learned something about how small our planet is and how interrelated is its weather.

How El Niño Works

During an El Niño, changes in major air pressure systems and prevailing winds push warm Western Pacific waters across the ocean toward the Americas. Sea level and surface temperatures rise in the Central and Eastern Pacific. Nutrient-rich feeding grounds disappear as they are overlaid with a large layer of nutrient-poor water.

Normally, the trade winds move that warm water eastward off the coast of South America and cause *upwellings*, the upward movement of deeper, colder water to the surface offshore. That colder water is richer in nutrients. Those nutrients feed the plankton—which feeds the fish—which keeps the big fish, the seals, and the fishermen happy.

In an El Niño, the warm water comes and
buries those cold upwellings under about
500 feet of warm water—and the fish, seals,
and fishermen go hungry. The area of
warm water in an El Niño is huge—about
one and a half times the size of the United

def•i•ni•tion

Upwellings are the upward
movement of deeper water to the
surface.

States. The 1997–1998 El Niño bore more energy than a million Hiroshima bombs.
Its energy spawned a whole suite of other weather phenomena, including floods,
droughts, tornadoes, and hurricanes. Does that sound familiar?

The History and Effects of El Niño

The effects of El Niño were documented in Peru as early as the Spanish conquest
in 1525 C.E., but researchers have found geological evidence of El Niño going back
13,000 years. Anthropologists believe that the Incas were aware of El Niño. That's
why they built their homes on the tops of hills. If they did build near the coast,
though, they kept back from the shore and stayed away from rivers. By the end of the
nineteenth century, Peruvian oceanographers were studying the phenomenon. At the
time, they thought El Niño's effects were limited to the west coast of South America.
It wasn't until the systematic studies of the International Geophysical Year that the
global extent of its impact was recognized.

In normal years in the equatorial regions of the Pacific, sunlight heats up the upper
layer of seawater around Australia and Indonesia. The evaporation off the water is part
of the monsoons that hit southeast Asia. The hot air loses its moisture as it rises into
the upper troposphere, where the winds there send it east toward South America. By
the time the air hits the coast, the winds are colder and dryer—and so they sink. The
sinking air creates a high-pressure zone, which is why there is dry desert along the
west coast of South America despite its proximity to the equator.

The trade winds push the sinking air—and the top layer of warm water with it—back
toward the Western Pacific. Along the coast of South America (but especially off the
coast of Ecuador and Peru), colder subsurface water wells up to replace the departing
warm water—bringing up a mass of nutrients for all the creatures in the sea. The end
result is that the coast off Peru is one of the most prolific fisheries in the world.

It Kills the Fish

But during an El Niño, the trade winds diminish or disappear. The warm water in
the Eastern Pacific just sits off South America soaking up the heat, blocking the cold

upwellings along with the feast of nutrients they provide for marine life. Fish that normally thrive off Ecuador and Peru head south in search of cooler, richer waters. The fishing gets good in Chile. The change is felt by marine life as far north as the Canadian border. During the 1997–1998 event, dorado (mahi mahi), a fish normally caught in tropical waters, were caught by sport fishermen in the waters off California. One fisherman caught a marlin off Washington State, and marlin are rarely seen north of Mexico.

When an El Niño arrives, Peruvian fishermen usually take a break, fix their nets, repair their boats, and spend time with their families. Mild events are generally tolerated; strong events can be economically devastating. The 1997–1998 El Niño put a serious dent in the Peruvian fishing economy.

It Floods the Cemeteries

As the ocean gets warmer, the water off Ecuador and Peru actually swells. During normal years, because of the westward-moving trade winds that push the water with them, the sea level is usually 18 inches higher off Indonesia than it is off South America. During an El Niño, the situation is reversed. As the water off South America gets hotter, it starts to evaporate. The moisture moves upward, creating a low-pressure zone off Ecuador and Peru. Torrential rains begin to fall in areas that aren't used to it.

In Peru during El Niño 1997–1998, mudslides—fed by the second-worst El Niño rains in Peruvian history and aggravated by loggers who clear-cut the forests of nearby hills—destroyed the entire village of Santa Teresa, killing 22 people. Three hundred bridges were washed out. To cross the San Francisco River, many of the people from surrounding towns had to commute by inner tube pulled by the local ferrymen. Flooding in the town of Mampuesto unearthed caskets from the local cemetery and floated skeletons through the streets.

Global Warnings

The runoff from the flooding in Peru during 1997–1998 was so severe that it created a lake in the coastal desert where there had been only arid wasteland before. The lake—90 miles long, 20 miles wide, and 10 feet deep—became for a time the second-largest lake in Peru. It served as a warning of the climate's ability to alter the landscape on a grand scale.

El Niño's Global Reach

Although El Niño involves only one fifth of the surface of the planet, it transforms weather around the world. El Niño currents not only influence the water and local weather, but the additional heat in the Eastern Pacific also feeds into the high atmosphere and creates a strong *jet stream* that races northward. It's always there, but it's much more powerful during an El Niño. The jet stream turns onto the North American continent and marches across to the Atlantic. Its effects are enormous.

def•i•ni•tion _____

Jet streams are strong ribbons of horizontal winds found 6 to 10 miles above the earth in the area between the troposphere (the lower layer of the atmosphere) and the stratosphere above it.

Mudslides in California

California seems to receive the brunt of the El Niño storms delivered to the United States. During the 1997–1998 El Niño, the local media kept doing stories on its coming impact—but as fall turned to winter and little rain fell, El Niño became the butt of nightly jokes on weather reports and late-night talk shows. But around February, the rains began to fall and the jokes stopped.

The first 24 days of February, Los Angeles received 13.5 inches of rain (close to its annual average). San Francisco's total for the season was 38.5 inches. These totals don't sound like much to someone from the Northwest or the Northeast, but you have to consider that the vegetation on the hills around California has evolved to hold much lower amounts of water. The rain saturated the ground and quickly exceeded the ability of the sparse vegetation to hold it back.

Much of the destruction created in California by El Niño came as mountains and coastline cliffs simply seemed to dissolve into debris slides or mudslides, taking homes, cars, livelihoods, and lives with them. The Amtrak rail line between Los Angeles and Seattle was severed by a mudslide in Santa Barbara. Ocean waves reached heights of 30 feet in spots along the coast. Helicopters regularly pulled surfers from the water. The capacity of the normal drainage systems, overflow channels, and riverbeds was quickly exceeded, and excess water became destructive. Two highway patrolmen died when their car was swept into a swollen river east of Santa Maria. Two college students were killed when a tree fell on their vehicle.

I remember being caught in a torrential downpour on the Los Angeles Freeway during one of the more severe of the season's rains. Visibility was suddenly cut to about 10 feet in the middle of the day as motorists slowed from 50 mph to 5 mph. All of us braced for the possible motorist who wouldn't bother to brake, or whose reflexes were unused to the conditions, to plow into the rear of one of our cars and set up a chain reaction.

Hurricanes Off Mexico

Warm Eastern Pacific waters set up prime conditions for hurricanes. Hurricane Linda grew out of the Eastern Pacific in September 1997. Her 185-mph winds made it one of the strongest Eastern Pacific storms ever recorded. The spiral arms of the hurricane lashed Mexico's western port cities. I took a bus down the coast from Puerto Escondido to Santa Cruz along the southern coast in the state of Oaxaca, Mexico, in the fall of 1998—a year after the hurricane—searching for sea turtles. I was amazed at how many concrete structures had been demolished by the storm. The sea turtle center at Puerto Angel, one of Mexico's prize natural exhibits, lay in ruins.

Although the Eastern Pacific delivered a number of intense hurricanes in 1997, the same was not true for the Atlantic. An El Niño-related tropical jet stream headed eastward across the Atlantic, effectively sheering the tops off the clouds before Atlantic hurricanes could form. During 1997, tropical storm activity in the Atlantic was diminished. The month of August passed without a single tropical storm developing over the Atlantic—the first time that had ever happened in the twentieth century.

Cherry Blossoms in D.C.

While the Southwest endured one of the wettest winters in recorded history, the winter of 1997 in the Northeast was the warmest on record. Cherry blossoms were blooming in January in Washington, D.C. The jet streams that travel 6 to 10 miles above the earth's surface shifted dramatically. The polar jet stream stayed further north over Canada than usual. The typical winter chill was reserved for those north of the U.S. border. It was a year of no winter for the mid-Atlantic region. Some estimates put the savings in heating oil for that year at $5 billion.

Ice Storms in New England

But winter didn't disappear; it just shifted northward. Quebec and northern New England were bashed with one of the worst ice storms in their history. Ice storms, you remember, occur when warm air flows over a freezing surface. Southern New England to northern Georgia are the normal victims of this type of storm. But during the big El Niño winter, the pattern moved northward and a heavy layer of ice accumulated during a storm in January. Ice felled enormous power lines and trees. Quebec and northern New England were brought to a standstill. It was one of the biggest ice storms ever to strike the East in recorded history.

Tornadoes and Floods in the South

Vicious storms pummeled the southern states. Snow fell in Guadalajara, Mexico, in the winter of 1997–1998 for the first time since 1881. In December 1997, Roswell, New Mexico, experienced its worst one-month snowfall since 1893. From Florida to Texas, tornadoes and floods were relentless. During the late evening of February 22 and early morning of February 23, 1998, a series of tornadoes ripped across central Florida. At least one of the tornadoes reached an estimated rotational wind speed of 260 mph at the center. The tornadoes destroyed 800 residences and left another 700 uninhabitable. Damages exceeded $60 million. Forty-two people were killed, making it the deadliest outbreak of tornadoes in the state's history.

Droughts in Asia

El Niño moves the rains that would normally fall on Indonesia, Australia, and India to the Eastern Pacific and produces drought in the Western Pacific. According to historical records, 600,000 people died in India from the epic droughts that accompanied the El Niño of 1789–1793.

Forest Fires in Sumatra

During the 1997–1998 El Niño, forest fires burned furiously in Sumatra, Borneo, and Malaysia, forcing drivers to use their headlights at midday. The smoke traveled thousands of miles to the west into the normally crystal-clean air of the Maldive Islands, where the visibility was limited to half a mile at times.

The Fall of the British Empire

One historian argues that El Niño could have contributed to the fall of the British Empire. According to Mike Davis in his book *Late Victorian Holocausts: El Niño Famines and the Making of the Third World* (Verso Books, 2001), the oscillations of the giant Pacific current and the revolutions of the early twentieth century are linked. Davis recounts how in the late nineteenth and early twentieth centuries a number of El Niño–related events led to famines in India, Brazil, and China. But famines were so ineptly handled by British imperialists that, according to Davis, they resulted in the deaths of 50 million people. Davis argues that imperialism had already robbed these formerly prosperous areas of their resources—and when calamity hit, the people had no fallback. When the imperial government didn't fill the void, its fall from power began.

> **Global Warnings** _____
>
> Asia, Indonesia, and Japan were hit by famine and fire during the 1997–1998 El
> Niño to the extent that it stretched thin the economic safety nets that many of these
> countries had during the late '90s. Although it was certainly not the only cause, many
> economists believe it was one reason for the great Asian Stock Market Crash of 1997.
> If global warming predictions are correct, what economic hardships might a rash of
> such climate-related events produce in the twenty-first century?

El Niño and Nature

The warm-water buildup off Ecuador and Peru is fatal not only to the fish, but to other marine life as well. In 1982–1983 during an El Niño, 25 percent of the adult seal and sea lion population perished, as did most of their pups, along the coast of Peru. During an El Niño event, the warmer-than-normal water sends tropical fish northward, but off the West Coast of the United States, those fish don't take up the void created by the disappearance of normal prey species.

Pacific sea birds are particularly susceptible to El Niños because coastal upwellings are their primary source of food. During an El Niño, fish die or go elsewhere in search of food. Birds do the same. The 1982–1983 El Niño event killed more than 90 percent of the adult sooty tern population of 14 million terns on Christmas Island.

The 1997–1998 El Niño devastated the California sea lion population in the islands off California and Baja California. Nursing females usually find plentiful squid and small fish near the surface, but the prey swam deep in search of cooler waters—out of sea lion diving range. In non-El Niño years, pups have a rough time—with mortality usually reaching 25 percent. But in 1997–1998, pup mortality reached 70 percent.

The heat that cooked the waters off Australia took a toll on marine life there as well. Along Australia's Great Barrier Reef, the water temperature got as high as 88°F (31°C). The coral polyps along some sections of the reef were bleached white. Most areas recovered, but along an 18-mile stretch off western Australia, the coral eventually died.

El Niño and Disease

El Niño brought more water to Peru, and that brought more disease. Water pooled in the eastern desert, and mosquitoes thrived. Those mosquitoes were responsible for the spread of malaria—some 30,000 cases in the Piura region alone, three times the average.

Bartonelosis, a bacterial disease transmitted by the bites of sand fleas, rampaged through cities on the western slopes of the Andes. In the chronic form of the disease, patients get long-lasting, blood-filled, wartlike sores on or under their skin, and the disease can be fatal in 40 percent of untreated cases.

def•i•ni•tion

Bartonelosis is a bacterial disease transmitted by sand fleas that causes wartlike sores on the skin.

Residents of Cuzco, a popular tourist town in the Andes near the Incan ruins of Machu Picchu, had never suffered outbreaks of bartonelosis until the 1997–1998 El Niño. Data from satellites that monitor the temperature of the Pacific show that the ocean began to warm about two to three months before the disease hit.

These weren't the only problems. Between January and May of 1998, 1,039 clinical cases of cholera were reported in Lima, Peru. The outbreak of that disease also coincided with the increase in temperature.

La Niña, the Wicked Sister

La Niña is the wicked sister of El Niño. During a La Niña, an abnormal cooling in the Eastern Pacific produces conditions more or less the opposite of El Niño. As with El Niño, the effects are most pronounced from December through March. During a La Niña, westbound trade winds are intense. That drives more than the normal amount of warm surface water westward. More than the normal amount of deep, chilly water rises up off the coast of Peru and Ecuador, producing a cold tongue-shaped section of water that extends from South America all the way to Samoa.

With all that heat moving west, the rising moisture from the Western Pacific creates such an intense low-pressure zone that it affects the monsoons off southeast Asia, drenches parts of Australia, and produces rains as far west as southern Africa.

The polar jet stream, which in an El Niño stays high in Canada, moves farther south during a La Niña—driving frigid air deep into the United States. Freezing winters visit the Northwest and upper Midwest, less rain falls in the Gulf and southeastern states, and drought is common in the southwest desert.

Global Warnings

The 1998 La Niña hurricane season was the deadliest in the last two centuries. Hurricane Mitch killed 9,000 in Central America. This compares to Katrina, which killed 1,836 people.

During an El Niño, a tropical jet stream pushes across the Atlantic—shearing off the tops of hurricanes approaching the East Coast. But during a La Niña, hurricanes in the Atlantic encounter no wind resistance and are twice as likely to strike the United States.

Keeping Track

The 1997–1998 El Niño marked the first time in human history that meteorologists were able to predict abnormal flooding and droughts months in advance, allowing time for threatened populations to prepare. The U.S. National Oceanic and Atmospheric Administration (NOAA) first announced the possibility of an El Niño as early as April 1997. Australia and Japan followed in May. By summer, the details of those predictions were available to countries all over the world.

In Peru, warnings reduced the number of tragedies. Grass grew on lands that were normally barren, so the cows got fatter. Rice and bean fields could be planted in areas normally too dry for those crops. Fishermen outfitted their boats for shrimp harvests, which the normally cool waters off Peru don't usually support. Even in some of the poorer regions of Peru, constructing storm drains, stockpiling food, and preparing for disease saved hundreds of lives.

Making those predictions involves a Herculean effort. One of the biggest helps has been the TAO (tropical atmosphere/ocean) array of 70 moored buoys that span the equatorial Pacific. Completed in 1994, the TOA buoys monitor water temperature from the surface down to 1,600 feet. The data collected by the buoys is transmitted to polar-orbiting satellites and then to NOAA's Pacific Marine Environmental Laboratory in Seattle.

Normally, predictions are made on a statistical basis. Data collected from previous El Niños, for instance, may show that lower barometric pressure and higher sea-surface temperatures in Tahiti usually mean more rain for Ecuador and less rain for northern Brazil. By determining what conditions led to past El Niños, scientists can be on the lookout for similar matchups in the future. But this kind of statistically driven approach doesn't tell scientists the mechanics involved.

With the advent of supercomputers, scientists are now able to input the data as well as fundamental laws of oceanic and atmospheric physics into simulated worlds and see how the weather changes over time. Scientists build these models based on present situations, inputting the data and seeing whether the models can produce a world that scientists can observe. When the models are refined enough to do that, scientists put them to work predicting the future. With these models, scientists were able to predict the 1997–1998 El Niño almost a year in advance.

The Big Switch

Scientists believe that El Niños have grown more frequent and more intense during the last century. During the twentieth century and into the twenty-first, there were about 23 El Niños and 15 La Niñas. The four strongest occurred after 1980.

Is El Niño changing? The IPCC reports that the frequency and intensity of El Niño/ La Niña events have increased since 1976 compared with the previous 100 years. The warm or El Niño phase of this team of weather phenomena has been relatively more frequent, persistent, and intense.

Global Warming

The link between El Niños and global warming is less than certain, but El Niños and global warming are about heat.. With an El Niño, heat is transported out of the tropics during the peak of the event—and this raises global temperatures. With global warming, there will be more heat available. The cycle might be shorter because the recharge time will be shorter.

With world temperatures higher, the water off Ecuador and Peru will take less time to reach critical temperatures. If global warming increases in the twenty-first century, scientists predict there might be both more frequent and more intense El Niños to contend with.

The Least You Need to Know

◆ El Niño events are caused when the warmer water and lower pressure normally found in the Western Pacific near Australia and Indonesia shift toward the Eastern Pacific near Ecuador and Peru.

◆ Along the normally dry equatorial coast of South America, El Niño events ruin offshore fishing and produce torrential onshore rains.

◆ El Niño events affect the weather worldwide.

◆ El Niños have increased in frequency and intensity since the mid-1970s.

◆ Scientists predict that if global warming increases, so will the occurrence of El Niños.

Chapter 11

Atmospheric Variables

In This Chapter

- ◆ Other factors that might influence global warming
- ◆ Earth's erratic orbit
- ◆ Sunspots and droughts
- ◆ Hot times in the city

Despite all the careful calculations of well-meaning scientists, we have to be honest. The case for global warming, if we exclude the carbon dioxide evidence (a big exclusion), is based on globally averaged surface temperature rises of 1.35°F (.74°C) over the last 100 years. That's not a lot. Is it possible that the causes of this rise could be related to something else besides greenhouse gases? A variety of astronomical, atmospheric, and geological effects have driven climate change over the history of the earth. In this chapter, I'll play devil's advocate and look at some of the things that could be changing our weather right now—things that in fact have changed our weather in the past.

Continental Drift

When scientists discovered evidence of an ice age that occurred 250 million years ago in present-day Africa, Australia, South America, and India, they were baffled. These were equatorial countries. What was evidence of ice-age glaciers doing in those areas? It wasn't until the theories of *plate tectonics* and continental drift came along that these ice-age remnants began to make sense. According to the theory of plate tectonics, the earth's crust is made up of a number of small pieces or plates that are floating on a partially molten layer.

Pangea—the super continent.

def•i•ni•tion

Plate tectonics is the theory that the earth is made up of plates that move around, bump, and collide with each other in ways that create some of the earth's most spectacular landforms.

Scientists who formulated this theory believed that these continents were part of a giant supercontinent called Pangea, which began to break into pieces and drift apart about 200 million years ago. In fact, if you look at the landmasses, they seem to go together like pieces in a puzzle. Pangea is probably not the only supercontinent that ever existed, but it's the one we know the most about.

Scientists believe that when this landmass broke up, the various pieces drifted apart on their own plates—eventually ending up where they are today. That explains how Africa, Australia, and India—which we find in subtropical latitudes today—could possess ice-age evidence. They were once found much farther south, near the Antarctic.

We now believe that continents drifting past or bumping up against each other are responsible for many of Earth's landmass forms, including its tall mountains, coastal bays, peninsulas, and islands. The Himalayan Mountains were created when the floating landmass of India plowed into southern Asia. These drifting continents must have also had a drastic effect on ocean currents as well as the atmosphere—although in different ways, of course.

Without the Himalayas and the Tibetan Plateau, India's monsoons might still be active—but they would be a lot less intense. But continental drift occurs at a snail's pace—a prehistoric snail at that. Continents might move two inches a year (often less). Their effect on weather is measured in millions of years, not centuries. Continental drift therefore does not fit into the present emergence of global warming.

Asteroids and Dinosaurs

How about asteroids? Scientists believe asteroids have made big changes in the past. The theory that a big one impacted Earth about 66.4 million years ago and led to the extinction of the dinosaurs was first advanced by scientists who discovered a high concentration of the rare metal iridium in rock strata near Gubbio, Italy. When dated, the iridium-laced rocks coincided with the end of the dinosaur era. Further evidence of the iridium/dinosaur link was found in a number of sites located in Denmark. Scientists found it in ocean floor sediments in the Pacific and Caribbean. To date the iridium layer, as it is called, has been identified in more than 100 places around the globe. The only thing that could account for the iridium is material from an asteroid, which would have a much higher level of this element than the rocks found on Earth.

Scientists attributed this iridium concentration to a collision between Earth and a huge asteroid. An impact explosion with a large asteroid would have ejected an enormous volume of earth and asteroid into the atmosphere. This could have produced a cloud of dust and solid particles that would have encircled Earth and blocked out sunlight for years. The loss of sunlight may have killed off the plants that the plant-eating dinosaurs ate, which would have wiped out the predator and scavenger dinosaurs as well.

Scientists now suspect that the giant Chicxulub crater in Mexico's Yucatan Peninsula could have been the asteroid that brought the dinosaurs down. However, they debate whether this impact was sufficient by itself to wipe out the enormous reptiles or whether other factors were involved. Some species of dinosaurs hung around quite a bit after the asteroid impact. It might be that to change the planet from *Jurassic Park* to what it is today required a number of changes—biological changes, climate changes, an asteroid impact, and maybe even a few volcanoes.

Asteroids can't explain the current level of global warming either, but they show you how climate can change quickly. Let's look at the earth's orbit for some possibilities.

The Earth Gets Around

The earth gets its heat from a fraction of the sun's heat. In fact, it receives less than two billionths of the sun's energy. The fact that the earth is moving in different ways affects how the sun's energy is received on Earth. The earth has basically two principal motions: rotation and revolution. It rotates as it spins on its axis, and it revolves around the sun. Both the rotation and the revolution affect climate.

The earth travels around the sun at approximately 70,200 mph (113,000 kph). Early humans thought the earth was still, or otherwise we'd feel the air rushing past us. The reason we don't feel the air rushing past us is because the air, held on by gravity, is going just as fast as the earth.

The Seasons

The seasons are created by a change in the length of daylight hours. As the seasons progress, the days get shorter or longer and the noon sun changes its altitude in the sky. In summer, the sun is high above the horizon as it makes its way across the sky. As winter approaches, the noon sun is lower in the sky and sunsets occur earlier each evening.

The seasonal variation in the angle of the sun affects the amount of solar radiation the earth receives. When the sun is overhead, its rays pass through a thickness of one atmosphere. But when the rays enter at an angle of 30 degrees, they pass through a thickness of two atmospheres; at 5 degrees, they pass through the equivalent of 11 atmospheres. The longer the path, the more chance for scattering, absorption, and reflection, which rob the sun's rays of their intensity.

The annual changes in the sun's angle and the length of the day are created by the earth's angle of orientation toward the sun. The earth's axis is not perpendicular to the plane of its orbit around the sun. It is tilted 23.5 degrees from perpendicular. Remember your globes? They are tilted on an angle, and that angle approximates the earth's attitude toward the sun. The tilt is called the *inclination of the axis*. But that inclination wobbles a little. Right now, the earth's axis is pointed at the North Star, Polaris. By about 14,000 c.e., it will be pointed at Vega, the new North Star. That's because of that wobble—which can affect the earth's orientation toward the sun (and, as a result, its climate).

On the day of the *summer solstice* (June 21), the north pole is inclined 23.5 degrees *toward* the sun. This is the beginning of summer in the northern hemisphere, and the longest day of the year. In the southern hemisphere, it's just the opposite—the beginning of winter and the shortest day of the year. Six months later on the day of the *winter solstice* (we're talking the northern hemisphere again, folks), the north pole is inclined 23.5 degrees *away* from the sun. This is the beginning of winter, the shortest day of the year—the day upon which the people in the northern hemisphere receive the least light and energy from the sun. The summer and winter solstices occur on opposite days in the southern hemisphere.

def•i•ni•tion

The **inclination of the axis** of the earth is a measure of the number of degrees that the axis is tilted away from the plane of its orbit around the sun. In the northern hemisphere, the **winter solstice** is December 21, when the north pole is inclined 23.5 degrees away from the sun. The **summer solstice** (again, we're talking the northern hemisphere) is on June 21, when the north pole is inclined 23.5 degrees toward the sun.

Both the orbit of the earth around the sun and the tilt of the axis vary in certain ways. These variances may have caused some of the climate changes in the past. Could they be creating the observed changes in the present?

Eccentricity

The earth moves around the sun, but its orbit isn't exactly circular. Rather than a perfect circle, like a basketball, sometimes its orbit is out-of-round, like a football (not that much, but you get the idea). The out-of-roundness or *eccentricity* of the orbit changes the distance from the earth to the sun at different times of the year. Currently, the earth is closest to the sun in January and farthest away in July.

However, we are in a period of low eccentricity—the earth's orbit around the sun is rounder—so the difference in the distance to the sun between January and July is slight (3 percent). That variance doesn't have much effect on the amount of the sun's energy that the earth receives. But when the eccentricity of the orbit is high, the difference can be as much as 20 percent—which can have a much bigger effect on the atmosphere.

The change in the eccentricity of the orbit from a circular one to one that's more elongated takes about 100,000 years. What is interesting is that the ice ages also peak about once every 100,000 years—and have for the last million years or so.

Doing the Wobble

The earth rotates on its axis, but it's not a perfect rotation. The earth, like a spinning top winding down, wobbles a little. The wobble changes the direction toward which the earth's axis is pointed. Right now, the north-end axis of the earth is pointed toward Polaris. When the Egyptian pyramids were built, the earth's axis pointed toward Thuban. In 14,000 years it will be pointed toward Vega. This wobble is called *precession*. One wobble takes about 19,000 to 23,000 years and causes each end of the earth's axis to circumscribe a small circle in space.

Tilt!

Remember the seasonal tilt that causes the difference between summer and winter? I mentioned it was 23.5 degrees. Well, that's what it is now, but it actually varies between about 22.5 and 24.5 degrees. When the tilt is less, the seasons will vary less—and glaciers have a better chance of expanding. Shorter, cooler summers will allow more snow to survive, and longer, milder winters would still be below freezing—but warmer air will produce more snow. The change in the tilt of the earth's axis is known as its *obliquity*. The changes in the earth's obliquity, from low obliquity to high obliquity and back, take about 41,000 years.

def•i•ni•tion

Eccentricity refers to the shape of the earth's orbit, which varies from round to oblong. **Precession** refers to the amount of wobble in the earth's axis. **Obliquity** refers to changes in the tilt of the earth's axis.

Ice Age Evidence

Let's look at all this evidence. The earth's eccentricity or "out-of-roundness" increases and decreases every 100,000 years. In addition, the earth's axis wobbles slowly—a process that happens every 19,000 to 23,000 years. And the tilt of the earth varies on a rotation that takes about 41,000 years. Although these orbital features don't affect the total sunshine received by the earth, they change where and during what season the sunshine is received.

So those changes in orbit and angle do affect the climate. But have they caused the amount of warming we've experienced in the last 150 years? Some scientists would have you think *Yes.* But the vast majority think not.

Climatoids

Scientists studying the bottom of the ocean are able to date the advances of the ice during the last ice age by measuring the chemical composition of the shells in each sediment layer. What they find is that over the last million years, ice grew for about 90,000 years, shrank for about 10,000 years, and repeated the process—with smaller disturbances about 19,000, 23,000, and 43,000 years apart. These statistics correspond to the orbit and tilt of the earth around the sun.

The Sun's Variables

Then there's the sun. Its output is variable. The problem is we weren't able to determine the variability of the sun's output until we were able to watch it from outside our atmosphere. We're doing that now from satellites, but we haven't been at it long enough. The only solar patterns we've looked at over time are sunspots.

Sunspots

Sunspots are some of the best studied of any of the features of the sun. They appear as dark areas simply because they are somewhat cooler than the surrounding gases. These spots are related to magnetic activity in the sun's outer layers, but the processes are not fully understood. Sunspots can appear in groups of two to more than 100. The average lifetime for a given sunspot is about one week. Sunspot cycles occur about every 11 years on average. When sunspot activity is high, the regions surrounding sunspots can have a number of solar phenomena including solar flares (bursts of high-energy particles) and prominences (huge solar clouds that erupt from the surface). During times of active sunspot cycles, the earth gets a bigger dose of solar energy.

People have been looking at sunspots for 2,000 years. Galileo paid particular attention to these little dark spots on the sun. We know that sunspots increase and decrease on a regular cycle, about once every 11 years. But in the historical records kept by scientists, there have been some periods with few or no sunspots. One of those periods was from 1645 to 1715. It turns out that this was also the coldest period of the Little Ice Age. It is referred to as the Maunder minimum.

Could this be the magic bullet? The second half of the twentieth century has been a period of relatively high sunspot activity.

Droughts and Tree Rings

In addition to the 11-year sunspot cycle, there is also another 22-year cycle in which the magnetic polarities of sunspot clusters reverse. Scientists who study the tree rings of the bristlecone pines in the White Mountains of California believe they have uncovered a sequence of droughts in the rings that matches the magnetic polarity reversals in the sunspots. The idea is controversial. Scientists can determine the amount of precipitation in each year by the width of the ring in the core of the tree. Wider rings represent happier trees, which in the arid west means periods of higher rainfall. It's not always that simple, however. "Happier" means different things to different trees in different areas. Furthermore, observations of other phenomena don't corroborate the magnetic polarity/drought theory.

Volcanic Ash

Scientists believe that volcanic eruptions can change the climate of the earth. Big, explosive eruptions can send huge quantities of gas and fine debris into the atmosphere. The bigger ones have sent debris high into the stratosphere, where it spreads around the globe and might remain for months or even years. The debris has the ability to filter out a portion of the sun's energy and lower the earth's temperature.

The most notable cold spell linked to a volcano was the one that followed the eruption of Mount Tambora in Indonesia in 1815. Many scientists attribute the abnormally cold spring and summer of 1816 to the clouds of volcanic debris put into the atmosphere by Mount Tambora. There were other, less-dramatic effects from Krakatoa in 1883 and Mount Agung in 1963. However, it wasn't until the eruptions of Mount St. Helens in Washington State in 1980 and El Chichòn in Mexico in 1982 that scientists really got to study the volcanic blasts with the use of sophisticated satellites and remote sensing instruments.

Mount St. Helens

Prior to 1980, Mount St. Helens was a steep, conical volcanic peak that rose 9,680 feet (2,950 meters) and had a snow-capped summit and even a few small glaciers. But on the morning of May 18, 1980, the entire north side of the summit came down—about 0.5 cubic mile (2 cubic kilometers) of rock and ice. An instant later, an enormous explosion of expanding steam and volcanic gas rocked the countryside. The gases formed a ground-hugging black cloud filled with hot, dense debris that raced over four major ridges and valleys up to 17 miles (28 kilometers) from the volcanic summit.

I visited the site after the eruption, and the devastation was awesome. For the first couple miles from the summit, everything had been obliterated. All you could see was a blanket of ash. A horseshoe-shaped crater 1.2 miles (2 kilometers) across and roughly 2,460 feet (750 meters) deep replaced the peak. But the most impressive thing was the blowdown zone, where huge virgin Douglas firs were snapped like matchsticks and lay on their sides, covered with ash. The U.S. Forest Service estimated that 10 million trees were felled by the blast.

Global Warnings

Although the destruction caused by the eruption of Mount St. Helens was enormous, the total magma displaced was only about .48 cubic mile (two cubic kilometers). The explosive eruption of Tambora Volcano in Indonesia in 1815 displaced about 12 to 24 cubic miles (50 to 100 cubic kilometers) of magma. That bested the 1883 eruption of Krakatoa, which displaced about 4.5 cubic miles (18 cubic kilometers). But civilization has never been tested by a cataclysm on the scale of the enormous eruption at Yellowstone about 2 million years ago, when more than 720 cubic miles (3,000 cubic kilometers) of boiling magma exploded into the air.

When Mount St. Helens erupted, there was instant speculation that it could have major effects on the climate. For a time, the large volume of volcanic ash that it emitted had significant effects both locally and regionally. But worldwide cooling was less than 0.2°F (0.1°C).

El Chichòn

However, studies following the eruption of El Chichòn showed a worldwide cooling effect on the order of .5 to .9°F (0.3 to 0.5°C). Why, if it was less explosive than Mount St. Helens, did it have a greater impact on global temperatures? The reason is because the material emitted by Mount St. Helens was largely fine ash that settled out relatively quickly. El Chichòn, on the other hand, spewed an estimated 40 times more sulfur-rich gases than Mount St. Helens. These clouds combined with moisture in the stratosphere to produce dense clouds of sulfuric acid droplets, which both absorbed and reflected solar radiation. Explosiveness alone is a poor instigator of climate change. For volcanism to impact the climate, you would have to have a whole bunch of volcanoes erupting over a relatively short period of time. Such an idea was once floated as the reason for the ice ages, but scientists have other ideas today.

Water and Carbon Dioxide

Everybody is looking at carbon dioxide as the principle villain for the amount of global warming we've seen thus far. But if an increase in greenhouse gases causes global warming, then the warming will create more evaporation. Water vapor is a powerful heat absorber, so the additional water vapor could compound the problem—more heat leading to more evaporation leading to more heat.

As the earth heats up, the cloud composition in the atmosphere could change dramatically. Low-lying clouds could evaporate during the daytime hours, and more high-flying cirrus clouds could form as surface heat causes the air to rise. Because low-lying clouds tend to reflect sunlight and higher clouds absorb heat radiation, the overall effect of these changes would be an increase in the amount of heat trapped in the atmosphere. On the other hand, the response of clouds to global warming is so uncertain that it could go the other way.

Urban Heat Island Effect

Maybe it's us. Maybe we are the ones causing the extra heat. At the beginning of the last century, only about 2 percent of the population lived in cities of more than 100,000 people. Today, it's more like 25 percent worldwide. In places like the United States, western Europe, and Japan, it's better than 50 percent. The *urban heat island effect* is one of the most studied and documented climatic effects on Earth. What this means is that temperatures within cities are generally higher than in outlying rural areas.

def•i•ni•tion

The **urban heat island effect** is a term used to describe the fact that urban environments are usually warmer than surrounding rural areas.

The effect is more pronounced at night, with minimums in the cities averaging 0.6 to 0.9°F (1.2 to 1.7°C) higher. But that's an average. On calm nights, the temperature difference between the city center and the countryside in Washington, D.C., can be as much as 20°F (11°C). The tall buildings, concrete, and asphalt of the city absorb and store greater quantities of solar radiation than do the vegetation and soil typical of rural areas. At night, the hard surfaces of the city release their heat more slowly than the earth. Heating, air conditioning, power generation, and transportation release a lot of waste heat. One study of a built-up portion of Manhattan found that during the winter, the quantity of waste heat was 2.5 times the amount of solar energy reaching the ground.

In addition, industry, cars, and heating appliances release a blanket of pollutants over the city, including particulates, water vapor, and carbon dioxide. These greenhouse gases make the city warmer than the surrounding territory. They also act as condensation nuclei, which encourage precipitation. Not only is it hotter in the city, it also rains more.

What effect does the urban heat island have on temperatures worldwide? It doesn't affect the world as a whole, but most temperature readings over the last 100 years have been taken in cities. The increase in temperature could simply be the effect of the cities on the temperature-takers. Determining accurate temperatures from past readings is problematic. It involves temperatures taken with different instruments at different times and with different standards. The 2007 IPCC report states that urban heat island effects are real—but local—and have a negligible influence on these values.

Could the total of all these variations, including continental drift, the earth's orbit, solar variation, volcanic ash, and clouds, be responsible for the 1.1°F (.6°C) change that scientists have recorded in the last century? According to the latest IPCC report, *Climate Change 2007*, "Warming of the climate system is unequivocal, as is now evident from observations of increases in global average air and ocean temperatures, widespread melting of snow and ice, and rising global average sea level."

Or, as William Schlesinger, president of the Institute of Ecosystem Studies in Millbrook, New York, says, "About as many scientists today believe that global warming is real and manmade as doctors believed that cigarettes were bad for your health in the late 1960s."

The Least You Need to Know

◆ The drifting of the continents on the surface of the earth and impacts by asteroids have altered the climate in the past.

◆ The out-of-roundness of the earth's orbit, the changing tilt of its axis, and the wobble in its spin have altered the advances and retreats of ice-age glaciers.

◆ Solar output, sunspots, and changes in magnetic polarity can affect our climate.

◆ Volcanic emissions can put a chill in your evening.

◆ Urban environments are hotter than rural areas, and this may have had an effect on past temperature readings.

◆ Most scientists feel that the evidence is strong for a human impact on our climate.

Chapter 12

Ratcheting Up the Equation

In This Chapter

♦ Things that could make it a lot worse

♦ The critical role of ocean currents

♦ Shutting down the currents

♦ A catastrophe in the making

In the previous chapter, I gave you some alternative explanations for the global warming that has occurred in the last century. So maybe things are not so dire. Maybe the scientists have it wrong. Maybe you can relax. Toss some fossil fuels into the fireplace, and take it easy. But before you slide into your easy chair, I want you to be aware of some natural triggers—which, just in case this global warming thing is for real—could make it a lot worse. Maybe you'll want to sit up and take notice.

Some scientists who've studied climate change over the ages believe that our climate is not a rheostat. In other words, it's not one of those dimmer switches that you turn down slowly. It's more like flipping a switch. As you apply ever-increasing amounts of pressure to the switch, it hits a point where it flips, the lights go off, it's dark, and you are in another state altogether. In this chapter, I'll look at some of the feedback mechanisms in the atmosphere that could ratchet up the pressure on that light switch and hasten the flip to another state. Of particular concern are the deep ocean currents—particularly the ones in the North Atlantic. We'll investigate what things could cause a change, what changes have happened in the past, and what we're liable to encounter in the future.

Feedback Mechanisms

We live in a climate system in which living things, land surface, ice, oceans, and the atmosphere are all linked up and interrelated. Within this complex system are various feedback mechanisms that help to balance each other. The body is a good example. If we exercise, we get hot—but then perspiration forms, which tends to cool us down. If you stopped sweating while exercising in warm temperatures, you could get heat stroke. Perspiration controls the buildup of heat. When this mechanism is turned off, it can unleash runaway processes. Snowball Earth was an example of one of these runaway processes that led to catastrophic results for all life on Earth.

If you recall, during Snowball Earth, the continents had drifted northward from the Antarctic to positions scattered around the equator, where it rained a lot. And the rain removed the carbon dioxide from the atmosphere, which reduced the greenhouse effect—and that caused the planet to get colder. Ice packs formed in the polar oceans, which reflected sunlight—driving temperatures even lower. A feedback process got started that proved unstoppable, and soon the planet was engulfed in ice.

The unusual configuration of the continents near the equator may have been the reason why Snowball Earth got going in the first place. When the continents were closer to the poles, the ice covered the land and stopped the carbon from being washed into the sea. This kept the carbon dioxide on land and in the atmosphere, which warmed up the place enough to prevent a runaway freeze. But when the continents were near the equator, there wasn't enough ice to stop the *carbon burial* process. A runaway freeze got started when the carbon burial process continued unchecked.

def•i•ni•tion

Carbon burial is the process whereby carbon from eroding rocks is washed to the sea where it ends up as carbonate sediments at the bottom of the sea. Sea life also adds carbon to these sediments.

Albedo

Now, let's look at another of these feedback processes that involves albedo. How well sunshine is reflected back from a given surface is what we call the albedo of that surface. If you had 0 percent albedo, that means none of the sunshine is reflected back, the surface absorbs it all, and temperatures get hotter. However, 100 percent albedo means that all the sunshine is reflected back, none of the heat is absorbed, and the surface stays cold. Black surfaces have 0 percent albedo; white surfaces have 100 percent or thereabouts.

Climatoids

Snow and ice mostly reflect heat, but the rainforest mostly absorbs it (due to different albedos). Thus, the removal of the rainforests might have the short-term effect of decreasing the heat absorbed at the surface, but the release of carbon dioxide from decaying or burning timber will have a net effect of more warming.

Clouds—particularly low, thick ones—reflect a lot of sunlight and therefore have a high albedo. High-altitude clouds tend to absorb heat. Dust and particulates influence clouds, and an increase in these elements can make it easier for low clouds to form—which might reflect the sun's energy backward.

The increase of sulfates in the air after World War II increased the clouds, which some scientists feel might have caused a mid-century cooling in an otherwise warming century. Snow and ice also have high albedos and thus reflect a lot of sunlight. The accumulation of snow and ice during Snowball Earth reflected sunlight, which led to the formation of more snow and ice.

The Bogs Are Brewing

English scientists report that peat bogs across the northern latitudes could be another problem. Peat bogs in Europe, Siberia, and North America may be decomposing and releasing carbon dioxide into the atmosphere. Peat bogs in the upper latitudes hold some 450 billion tons of carbon, which is the equivalent of 70 years of industrial emissions.

Scientists have recorded a 65 percent increase in the release of carbon from British peat bogs in the last 12 years. Scientists believe that much of the carbon locked up in soils for thousands of years will end up as carbon dioxide in the atmosphere. This will enhance the greenhouse effect and further accelerate the decomposition of the bogs.

Ice Melt Kills

Researchers agree that greenhouse warming will cause a substantial rise in sea level due to the acceleration of the ice melt on land and at the poles. Some scientists, however, contend that the increase in moisture caused by the warming will result in an increase in snowfall over Antarctica. The IPCC currently believes such a counterbalance may be occurring.

Global Warnings

If all the ice in the world were to melt, it would cause a rise in sea level equal to 230 feet (70 meters)—enough to drown most of the major ports of the world.

But the worst-case scenario, according to some scientists, is the breakup of the West Antarctic sea ice. This mass of ice is a mile thick and spreads over an area the size of India. It rests near the sea floor like an enormous grounded ship. As the climate becomes warmer, the West Antarctic sea ice could break up and possibly melt. It would take a while, but once started there'd likely be no stopping it. The result could be a rise in sea level of 10 feet (3.4 meters) worldwide.

Ocean Currents

The earth's climate is forever trying to rectify the heat imbalance between the tropics and the poles by shuffling heat from the equator to the polar regions. We talked about the part that the atmosphere plays in this heat exchange, but the atmosphere is only responsible for about half the heat. The other half is transported through the oceans. The water around the equator is heated more than any other ocean water, and the ocean moves that heat toward the poles to even out the heat budget.

Most of the surface waters of the ocean are moved by friction created by winds. They push on the surface of the ocean and eventually produce large currents that move water clear across the ocean. The trade winds push the Pacific waters from Peru to Australia. They also move Atlantic waters from Africa across to South America. The Gulf Stream of the Atlantic and the Kuroshio Current of the Pacific are some of the major currents created by wind.

Not only does heat and circulation play a role, but salt is a factor as well. Salty water is denser or heavier than normal water, and this is a factor in the great conveyor belt–like circulation of the deep oceans. If you place dense water over less-dense water, the less-dense water will sink—creating a current even without the help of the wind. You can make water more dense or heavier by making it colder or saltier. Water becomes saltier when evaporation removes the moisture but leaves the salt behind.

The Global Ocean Conveyor Belt

Surface currents aren't the only things going on in the sea. There are also global ocean currents that circulate on the surface and in the deep sea, connecting the major oceans on the planet. The scientific name for this conveyor belt–like circulation is the *Thermohaline Circulation* (*THC*) or the Global Ocean Conveyor Belt. In this current, warm, salty Atlantic water near the equator moves north toward Greenland and Labrador, where it cools and sinks. The sinking occurs in a couple of places in the North Atlantic, diving more than a mile deep before the water flips around and heads south back down the Atlantic.

The water flows south and swings around the tip of Africa, rising to the surface in the Indian and Pacific Oceans—as well as in certain spots around the coast of Antarctica—before it heads back toward the Atlantic and north again.

def•i•ni•tion

The **Thermohaline Circulation (THC)**, also called the Global Ocean Conveyor Belt, is a global ocean current that operates like a conveyor belt running through shallow and deep-sea waters to connect the major oceans of the planet.

The Critical Sinks

The Atlantic waters in this system that begin at the equator and travel north arrive at two relatively small spots: one east of Greenland and one east of Labrador. At this point, the water has grown heavy and dense enough from the cool air and the salt it brings with it to sink to the bottom.

The only other place where the waters descend along this conveyor belt is off the coast of Antarctica. There, the freezing of the water along the shelf takes up water but not salt, leaving a bath of salty water for the deep waters that rise up to the surface. Those waters linger in the salt, lose a little heat, and make another dive again—continuing their long journey. There are no such sinks in the Pacific Ocean because the water is simply not salty enough (a prerequisite if the water is going to sink to the bottom).

The greatest effects on the atmosphere from the THC are made in the North Atlantic, where the heat from the tropics is transferred north and released into the atmosphere before it sinks. Lacking those currents, much of Europe's climate would more closely resemble that of central Canada than its comparatively mild climate currently.

Jamming the Works

So how do you stop the deep ocean conveyor belt—the THC? Richard Alley, in his book *Two-Mile Time Machine*, compares the deep ocean conveyor belt to a grocery store conveyor belt and asks what you would do to stop that conveyor belt. His suggestion is to grab a fork and stick it into the small gap where the grocery store conveyor belt goes down. The conveyor belt will pull the fork down with it and get stuck.

Salt, or the absence of it, works like a fork to clog up the conveyor belt at the point in the North Atlantic where the water sinks to the bottom. The flow of water from the tropics to the North Atlantic exists within a delicate salt balance. The tropical ocean sends warm, salty water to the North Atlantic, but at the same time it sends water vapor north—which falls out as rain and snow. That moisture falls on the ocean or flows to it from land and decreases the Atlantic's salinity. If enough freshwater is added to the mix, then when the currents arrive in the North Atlantic, they may not be salty enough to displace the water beneath them.

Remember, water needs to be cool and salty to become dense or heavy enough so that it sinks to the bottom and continues the motion of the conveyor belt. As the water sinks to the bottom, it draws up more water from the equator—bringing its gifts of heat and moisture.

So how's global warming going to affect the THC? It will not only increase the warmth of the water, but even more importantly it will also increase the amount of precipitation. This freshwater is dangerous because it will, in general, decrease the percentage of salt in the North Atlantic by diluting it with freshwater.

Then what? Well, if the water gets fresh enough, the sinking will slow or stop—and water that arrives from the tropics, cooled by the atmosphere, will simply pool and possibly freeze. The ice will increase the albedo, and it will get even colder. Global warming will not be warm for everyone.

What If the Conveyor Belt Stops?

If the THC were to stop, it would also stop the transfer of heat from the tropics to the northern latitudes of the North Atlantic. Some scientists suggest that the heat given off by this current is the reason why northern Europe is at the same latitude as Alaska yet is much warmer. This current influences the eastern or European side of the Atlantic more than the western or New England side.

There are, however, other influences. The prevailing westerly winds blow over the ocean and spread its warmth over the west coasts of the continents. The jet stream also brings Arctic air down over New England and the Asian coast of the Pacific but brings warmer breezes to the U.S. Pacific Northwest and Europe.

Having said that, the THC is an important influence in the climate of the North Atlantic. Complete shutdown of the THC would remove an enormous amount of heat that is transferred from the tropics to the North Atlantic. If the THC were shut down, it could create a net loss to the region as high as 7.2°F (4°C). This could increase the amount of sea ice, and that could lead to even more cooling.

Global Warnings

The shutoff of the THC in the North Atlantic could lower the temperature of the region by about 7.2°F (4°C). That's the difference between this age and the ice age. Global warming might signal hot times for the rest of the world but a deep freeze for the North Atlantic.

Dead Roses for England

Great Britain is downwind of where the conveyor belt dumps its gift of warmth and moisture. Subsequently, British citizens grow roses at the same latitude where polar bears live near the Hudson Bay. The warm, salty water that enters the North Atlantic must get rid of its heat to make the dive, and England and much of northern Europe gets to share in that bounty.

Climatoids

Wind blowing over warmer waters and carrying heat and increased moisture is familiar not only to people who live in Great Britain but also to residents of Buffalo, New York, and Erie, Pennsylvania. In these places, subzero winds that blow down from the Arctic in the winter encounter the warm waters of the Great Lakes and pick up moisture and heat. They then dump their gift on communities below the lakes. The result is that while people out on the Great Plains sometimes suffer –40° weather, people in the snow belt under the Great Lakes are covered with a blanket of white that is just a little below freezing.

In models of the ocean and atmosphere, if you add enough freshwater to the mix of the conveyor belt that feeds the North Atlantic, the water there sinks more slowly. If the sinking slows enough and precipitation continues, the sinking could stop. Once the conveyor belt is shut off, its transport of heat to the north is terminated. The cool water pools on the surface and may freeze in the winter. This reflects heat and makes it even cooler. So once it's shut off, feedback mechanisms are engaged—which make it really difficult for the circulation to get going again.

According to Lynn Talley, professor and research oceanographer at the Scripps Institution of Oceanography at the University of California in San Diego, the results of a shutdown would be catastrophic. Eventually, enough ice would freeze on the land to remove the freshwater that caused the dilution—but it would take about 1,000 years. That means that the Royal National Rose Society in Great Britain could be on hiatus for an entire millennium.

It would also decrease the rising of summer-heated air from Africa, which flows into Asia and drives its summer monsoons. A cold north and a drier Africa and Asia would decrease the wetlands, which would decrease the production of methane (which, as you recall, is another greenhouse gas). Reduction of greenhouse gases makes it even colder. Colder air typically carries less water, so that much of northern Europe could be subject to both drought and bitter cold.

Permanent El Niños

I know what you're saying. Wait a minute. You thought we were talking about global warming. How do we get from global warming to ice-age Europe? What you really want to know is where would all that heat go? Like ducks flying south for the winter, the millennial winter that could result from shutting down the global ocean conveyor belt—the THC—would drive the heat south across the border.

Parts of the global ocean conveyor belt have shut down a number of times going back a million years or more—not from global warming but from other factors that upset the delicate warm, salty water mix. When it has shut off, parts of the northern hemisphere have gotten colder while parts of the southern hemisphere have become warmer—even during the ice ages.

So what could that heat do? All the things that extra heat does. It would increase precipitation along the equator and decrease precipitation in drier inland areas. If the north decreases its temperature by 7.2°F (4°C), then the world would seek equilibrium and the south could increase its temperature that much on average. The South Atlantic, which would have no place to send its warmth, might sit below the equator and get hotter. This could have momentous effects on tropical storms, including hurricanes.

If we take 7.2°F (4°C) from the north and give it to the south, that amount of increase might be sufficient to kick in an El Niño. Only … this El Niño could be permanent.

The Younger Dryas Event

If this stuff sounds all too fantastic or hard to believe, consider that such a freshwater dilution occurred just prior to the Younger Dryas event. Although the cause was different, the mechanics were similar. The ice that advanced down the North American continent during the last ice age had the effect of blocking the rivers that flowed north and east into the Atlantic. The rivers that flow from the Great Lakes were so clogged that the lakes rose to a point where they overflowed into the Mississippi River and eventually ended up in the Gulf of Mexico.

But when the ice began to melt, the eastern outlets down the Susquehanna, the Hudson, and finally the St. Lawrence rivers gradually opened. Water breaking through an ice dam rapidly opens an ever-larger hole in the ice. Ice dams are recipes for catastrophe, and this was no exception. A huge flood raced down the St. Lawrence into the sea as the Great Lakes broke through the ice and sought out a new level. This enormous dumping of freshwater into the North Atlantic reduced the salinity of the poleward-moving currents and began to turn off the conveyor belt. The Gulf Stream virtually stopped. Western Europe no longer received all that heat from the evaporating Gulf Stream. Although it was not the only time this has happened, such a huge flood did occur down the St. Lawrence River just before the Younger Dryas—and scientists believe they are related.

Remember the Younger Dryas from Chapter 6? Ice core evidence from Greenland shows that windblown dust and sea salt increased by a factor of 3 to 7 times. In Norway, mean July temperatures were about 12 to 16°F (7 to 9°C) colder than today. In Spain, the July temperatures were as much as 14 °F (8°C) colder than today.

The Younger Dryas affected North America as well. It threw us back into the ice ages for more than 1,000 years. The greatest catastrophe of the Younger Dryas event may have been the extinction of the great ice-age mammals. Although early human hunters definitely played their part by overhunting, the mastodons, mammoths, horses, and saber-toothed tigers—which once wandered the North American continent—became extinct during the Younger Dryas.

There might be switches that have catastrophic effects as extreme as the global ocean conveyor belt (the THC). It may just be one of the major climate switches that scientists have uncovered in their search to understand climate change in the past. According to scientists studying this phenomenon, flipping the THC on and off doesn't seem to account for all the drastic climate changes that Earth has seen in the past.

The Deep-Sea Heat Sink

As if the THC wasn't enough to worry about, scientists claim there are other troubles brewing down there in the deep sea. The problem is that given the level of carbon dioxide in the atmosphere now, a number of Earth scientists calculate that Earth should be .9°F (.5°C) warmer than it is today. Temperature readings obtained from several hundred meters below the ocean's surface suggest that this extra heat could be could be lurking down there.

The idea is that as the Earth has warmed, a lot of the extra heat has gone into the ocean—and currents in the ocean may have transported it down into the deep sea, effectively removing it from the surface of our planet. If this continues, the temperatures in the deep sea could increase until they are in balance with the surface waters. Once they've reached that balance, the excess heat would no longer be drawn down to depths—and the ocean as an absorber of the excess could cease to exist. After that, things would get a lot hotter real quick.

The Gamble

We are engaged in a radical experiment—a lethal gamble. We are on the way toward doubling the amount of carbon dioxide in the atmosphere, and scientists who study the atmospheric past tell us that such radical increases in this and other greenhouse gases in the atmosphere in the past have had wide and catastrophic results. Although the future looks grim, it may be that the global warming we've experienced won't continue and that the scientists are wrong. But … they may be right.

It's a lot like staring down the barrel of Dirty Harry's .44 magnum, wondering whether it's really loaded, and he's saying, "You've got to ask yourself one question: 'Do I feel lucky?' Well, do ya, punk?"

The Least You Need to Know

- Various triggers in the atmosphere could act as feedback mechanisms to increase or decrease the causes and effects of global warming.
- If the West Antarctic sea ice broke up and melted, it could raise global sea levels by 10 feet (3.4 meters).
- A current that operates like a huge conveyor belt connects most of the oceans of our planet.

◆ Global warming could decrease the salinity in the North Atlantic, which could turn off the global ocean conveyor belt.

◆ Scientists believe that the global ocean conveyor belt stopped just prior to the Younger Dryas event, which threw the world back into an ice age for more than 1,000 years.

◆ If global warming were to switch off the deep ocean conveyor belt again, it would be a catastrophe.

Part **4**

The Crystal Ball

So I've looked into the past and taught you a little about our weather and climate. Now I'm going to look into the future. I'll show you what the Intergovernmental Panel on Climate Change, a project of the World Meteorological Organization and the United Nations Environment Program, predicts global warming may do to Africa, Asia, Europe, and Latin America. How will it affect forests, agriculture, and tourism? Will it sink low-lying island nations?

Then you'll see what the U.S. Environmental Protection Agency predicts will happen in North America. What effects will it have on ranching, fisheries, and human health? Will Florida sink beneath the sea? I'll take a look at what the National Wildlife Federation predicts might happen to our wildlife, plants, and national parks.

I'll also investigate climate models and how they are used to foretell the future. If scientists can't predict the weather more than two weeks out, how do they propose to predict the climate over the next 100 years?

"So, um, Noah—you thinkin' about takin' a few other guys with you?
I mean, I might be available."

Chapter 13

The Global Picture

In This Chapter

- ◆ Tough times for Africa
- ◆ Asia gets drenched
- ◆ Latin America gets sick
- ◆ The islands go down with the ship

I've discussed the worldwide effects of global warming, but what do scientists think global warming will do to specific regions? Human populations and natural systems differ radically in their vulnerabilities. An increase in rain will cause an increase in the number of mosquitoes and the number of mosquitoes carrying disease, but each region will be susceptible to different diseases and even different mosquitoes. In some areas, it will rain a lot more; in other areas, it will rain a lot less. Global warming will further exacerbate extremes of all types of weather.

Regional differences will depend upon conditions and locations. Do you live in a desert? A swamp? A ski resort? A tropical island? All of these different areas will be affected by current weather, the change that global warming will make to it, and the resources at their disposal. A ski resort may be able to live with a little less snow, but a tropical island won't survive a 3-foot (91 cm) sea level rise if it's only two feet (60 cm) above sea level in the first place. Some areas will get wetter. Elsewhere, a desert may rise. In this chapter, you'll see what could happen to different regions around the world if the climate changes in the next century as much as some scientists predict.

Africa

As long as the infrastructure holds up, wealthy individuals and countries will be able to adapt to most short-term changes—whether it means buying another air conditioner or switching to warmer-weather crops. Africa, however, will suffer. Some areas like southern Sudan have already seen major changes. Lake Chad in southwest Chad, bordering Nigeria, once as large as Lake Erie, has shrunk to one twentieth of its original size. This has touched off famines throughout the region.

Africa is a diverse region with different climates, geography, culture, and economies. It is predominantly tropical, hot, and dry. Tropical rainforests exist in the west and central portions along the equator. But as you go north or south, the land becomes more arid. Most of the population lives in the sub-humid areas, north and south of the tropical forest, and in the semi-arid areas a little north and south. Different regions will have different problems.

Tough Times for Agriculture

More than half of the people of Africa live in rural communities and are dependent on local crops harvested from nearby fields. Irrigation is inadequate, and many farmers have to rely on natural rainfall. While the rest of the world keeps figuring ways to grow more food, Africa's ability to feed itself keeps declining. Over the past two decades, Africa has suffered widespread famine and malnutrition. Its nations frequently have to rely on emergency food.

In the drought years of the mid-1980s, African nations ate about 50 percent more than they could grow. In the mid-1990s, consumption exceeded domestic production by 30 percent. One third of all Africans eat fewer than 2,000 calories a day, which most residents of industrialized nations would call a diet. The continent needs to increase its agricultural output by 4 percent a year just to meet basic needs. But to achieve that, it may have to double its farmlands by 2050. And that will increase the pressure to cut down tropical forests.

Now add global warming to the picture. Frequent and prolonged droughts in savannah areas would devastate agriculture, promote famine, and place the population at peril. There has been chronic war in several regions of Africa for the last three decades. Africa simply can't cope with the added stress. Government agencies are underfunded and undermanned. In many instances, military coups, despotism, tribalism, corruption, and the stringent economic constraints of the international financial community have brought them to the brink. Africa's troubles would only be exacerbated by the additional burdens of global warming.

More Disease

Africa is susceptible to the climate changes created by El Niño. If the world's climate were to shift into a permanent El Niño, as some scientists predict, Africa would suffer. With the heavy 1997–1998 El Niño rains in Rwanda, the incidence of malaria increased by 337 percent. Flooding in arid areas could promote the growth of malaria-bearing mosquitoes. In the Sahel region, which has suffered from drought over the last 30 years, malaria has decreased as still-water breeding sites for mosquitoes have dried up. But bring in a flood and let the water pool up—and malaria could make a comeback. The cities of Nairobi, Kenya, and Harare, Zimbabwe, were formerly located above the altitude of mosquitoes. But as the world has warmed, mosquitoes have moved into these higher places.

Extreme weather brings disease. Floods contaminate public water supplies, and drought encourages unhygienic practices due to a shortage of water. Cholera was epidemic during the 1997–1998 El Niño. Africa then reported 80 percent of the total number of cholera cases worldwide. Meningitis, on the other hand, is associated with warming and reduced precipitation. So while the flooded parts of Africa may be suffering from cholera, the drought-stricken parts could come down with meningitis.

Global Warnings

Following the 1997–1998 El Niño, Rift Valley fever—a disease of domestic animals—broke out in East Africa and killed as much as 80 percent of the livestock. The disease was thought only to affect livestock and sometimes domestic animals but never people. But the El Niño Rift Valley fever outbreak spread to livestock owners and rangers. What other surprises await if climate changes set in on a more permanent basis?

Asia

Asia would have problems with global warming. It already has more than 60 percent of the world's population, its natural resources are stressed, and its ability to adapt is poor. Many Asian countries are dependent upon natural resources like water, forests, grasslands, and fisheries. Sulfate emissions from Asia's rising industrialization could dampen some of the effects of global warming by filling the atmosphere with tiny droplets of sulfuric acid that will reflect away excess sunlight. But the acid rain that sulfates produce eats away marble and is deadly to inland fisheries. Using one toxic substance to counter another doesn't make a lot of sense.

The Sea Rises

More than half of the continent's population, 1.7 billion people, lives in the coastal areas of Asia. In countries like Bangladesh, Nepal, the Philippines, Indonesia, and Vietnam, many river watersheds suffer badly from deforestation, indiscriminate land conversion, and soil erosion. Global warming would impact these ecosystems by accelerating sea level rise and more frequent and more intense storms.

Three million years ago, when the temperature was 3.5 to 5.5°F (2 to 3°C) warmer than today, sea level was 25 meters higher—more than enough to drown New York, London, Tokyo, and Shanghai. Low-lying coastal cities like Jakarta, Tokyo, Manila, and Bangkok would also be affected by sea level rise with its saltwater intrusion, increases in silt, and loss of land. There are at least 10 major deltas in the coastal zones of Asia with an area of more than 3,844 square miles (10,000 square kilometers) each. Low-level deltas are especially sensitive to rising seas and pounding surf. With resources stressed to the max, little ability to adapt, and weak economic structures, global warming could wreck tropical Asia.

Bangladesh could lose close to 11,600 square miles (30,000 square kilometers) of its delta regions based on projected sea level rises of 39 inches (100 centimeters), while Indonesia may lose 11,550 square miles (29,900 square kilometers) and Vietnam 15,400 square miles (40,000 square kilometers). Such a sea level rise would expose 13 percent of Bangladesh's population, 2 percent of Indonesia's, and 23 percent of Vietnam's. That's a total of 34 million people for all three countries.

Global Warnings

Sea level rise was part of the catastrophe of the 2004 Indian Ocean earthquake, which triggered a series of lethal tsunamis on December 26, 2004, that killed approximately 300,000 people. The initial surge was measured at a height of approximately 108 feet, making it the largest earthquake-generated tsunami in recorded history. A rise of 39 inches generated by global warming seems tame compared to this disaster. Though these two operate on very different timescales and have very different causes, global warming will change the sea level and, possibly, change exposure to tsunami surges.

Tough Times for Agriculture and Aquaculture

In general, things may get better for mid- and high-latitude regions where lengthening of the growing season and an increase in carbon dioxide could potentially help production of some crops. But for lower latitudes exposed to El Niño and tropical storms, it will get worse. Yields of several major Chinese crops are expected to decline. Acute water shortages combined with increased heat will be rough on wheat and rice production in India, even with the beneficial effect of more carbon dioxide. Crop diseases such as wheat scab, rice blast, and sheath and culm blight could spread in temperate and tropical Asia as the climate becomes warmer and wetter.

Asia produces 80 percent of the farmed fish, shrimp, and shellfish in the world. Wild stocks are overfished. Trawlers destroy the sea bottom. And coastal development and pollution further stress the habitat. El Niño events already affect plankton, sardines, and prey species that are vital to marine ecosystems. Conservation of marine resources is behind schedule, and global warming will only make that worse.

Australia and New Zealand

Australia and New Zealand cover a region that spans the tropics and the midlatitudes of the southern hemisphere. The region has a variety of climates and ecosystems including rain forests, deserts, alpine areas, and coral reefs. The surrounding oceans strongly influence both countries. Australia already has significant desert areas, and periodic droughts plague agriculture. New Zealand is smaller, more mountainous, more temperate, and might prove more resilient to global warming.

Increased carbon dioxide may initially boost agricultural output, but that could change as droughts and floods get more severe. Plus, climate change may help spread pests and diseases. If droughts increase, they will hit rural communities the hardest. Droughts will increase the chance of forest fires. Basically, the initial advantage of a little global warming and an increase of carbon dioxide in the atmosphere will be offset by the fire, pestilence, droughts, floods, and storms that will come with it.

A warming of just 1.8°F (1°C) could threaten the survival of plant species growing at the upper limit of their temperature ranges, particularly in the alpine regions in New Zealand and in southwestern Australia. Species that can't move or migrate because man has cleared the land, because the soil is different, or because there's a mountain or river in the way could become endangered or extinct.

Things could become catastrophic if temperatures climb 5.5°F (3°C) above present levels, which could happen long before the end of the century in the worst-case scenario of the IPCC's predictions. Certain plants will thrive as CO_2 rises—but at the expense of others. Tropical coral reefs, which have already experienced widespread bleaching, would be particularly hard hit as well as the rainbow of marine species associated with them.

Europe

Europe is a rich and stable region with well-developed economies, stable governments and political institutions, and good technical support systems. The urban environments will do better in a globally warmed world. The natural environments, which have been overutilized for centuries, won't. Most natural ecosystems in Europe are managed, fragmented, or under stress from pollution or other human impacts.

Europe has a total population of 720 million people. Although it does not have as many people as Asia, it has a lot less land. In fact, Europe has the highest population density of any other continent. It has the lowest birthrate and the most rapidly aging population of any other continent as well.

> **Hot Debates**
>
> Europeans in general seem more concerned about global warming than much of the rest of the world, possibly due to the fact that if the global ocean conveyor belt—the Thermohaline Circulation—shuts down, Europe could be the big loser.

Potential impacts vary substantially between northern and southern Europe. Water shortages are projected for the south, while floods will hit the north. More marginal and less wealthy areas will be less able to adapt.

Europe has seen vast changes in precipitation in the twentieth century. Rain and snow have increased over northern Europe by as much as 10 to 40 percent, while at the same time southern Europe has dried as much as 20 percent. The number of rainy days over Spain's southern coast and Pyrenees Mountains decreased by 50 and 30 percent, respectively, in just the last half of the twentieth century.

Forests

Most European forests are managed for timber, water, or recreation. This process has basically reduced or modified forests over much of the continent. Present-day forests often consist of trees that are different from those that would have occurred naturally. High summer temperatures and drought could increase insect populations and bring down those trees.

Fewer Attractions for Tourists

Higher incomes and more leisure time should expand the tourist industry in Europe, but global warming may change where the tourists go. The Mediterranean might lose a little of its popularity under increased summer heat, prolonged droughts, and forest fires. The coasts of Europe are a primary attraction for the tourist industry, but sea level rise might damage buildings, bridges, and breakwaters. Sea level rise could also spell disaster for places like Venice, Italy, which is sinking already.

Lots of long, hot, polluted summers might make European cities less attractive. The mountains might get a bigger share of the summer business, but ski resorts will suffer in winter because of reduced snow. If the Alps increase in temperature by 5.4°F (3°C), the snow line will rise by 984 feet (300 meters), the first snowfall will be delayed, and below 3,927 feet (1,200 meters) there will be no continuous snow cover. Resorts might turn more often to artificial snow, and Scotland's skiing industry could dry up altogether.

Latin America

The region of Latin America includes all of the continental countries of the Americas from southern Mexico to Chile and Argentina. Despite the fact that the annual population growth is declining from 1.68 percent in 2000 to an estimated .51 percent in 2050, that will still mean another 838 million mouths to feed. One of the major difficulties for this region is having enough food. Many countries already suffer from poverty, deforestation, and dwindling resources, and global warming will exacerbate these problems.

Storms

Flooding and landslides are common in Central America and southern Mexico. Strong weather and the inability of the land or the society to contend with it have produced numerous disasters. Latin America also has one of the greatest disparities of income in the world. The rich seem to get richer and the poor get poorer here, and the poor often bear the brunt of major storms and changes in climate. The rich build their haciendas on dry spots up on hills. The poor throw up shacks in dangerous river valleys that are particularly susceptible to flooding. Hurricane Mitch created some of its worst havoc in river valleys in El Salvador and Nicaragua, where raging waters wiped out entire shanty towns.

El Niño and La Niña typically affect the various regions of Latin America dramatically. Some of the most severe droughts in Mexico in recent decades have occurred during El Niño summers. Then the Pacific side of Central America gets less rain than normal while the Caribbean side gets more.

Whereas the enormous Andes Mountains get deluged in an El Niño, the Amazon Basin north to the Caribbean suffers from intense droughts and numerous forest fires. At the same time that the north of Brazil is burning, the south is under the same intense rains that are falling over the Andes. The principal danger with a globally warm Latin America is a permanent El Niño condition, with its extremes of weather and a reversal of conditions to essentially the opposite of what the economies, societies, and geography have become adjusted to. That means rain in normally dry climates and dry climates in normally wet ones.

Again, the effects will be highly variable. In 2005, while Hurricane Katrina was drowning the southern coast of the United States, the Amazon endured one of its biggest droughts in decades. In a globally warmer world, both hurricanes and droughts will be come more prevalent.

Bugs

Latin America is famous for its wildlife diversity. The tropical forests that surround the equator, shroud the Amazon basin, and run north through Central America into southern Mexico have some of the most varied plant and animal species in the world. The same goes for their bugs. I remember going to Corcovado, a tropical forest in Costa Rica, and having the honor of being one of the first Americans to visit after it had been declared a national park. For my stay, I was told to bring "nets"—so I packed a roll of cheesecloth, thinking that would do.

By the time I got to Corcovado, I had already traveled through Mexico, Guatemala, El Salvador, and Honduras, and my insect repellent was getting low. When I arrived there, I was shown to a large sleeping dormitory where there were no screens to keep out the bugs. Instead, everyone there had full nets hanging from the ceiling surrounding their bunks like something you'd see in an English movie about Africa.

Different species of bugs arrived throughout the night, and you could watch the change of the insect guard around the light the camp kept on outside the building. It was too hot to use blankets. I tried to surround myself with cheesecloth, but even that got hot, and I would kick it off in my sleep and the bugs would start eating the exposed flesh. When I woke up the first morning, I counted the bug bites on my left arm, which had dangled outside the cheesecloth. From my hand to my elbow were over 40 bites, and around my elbow there were so many I couldn't differentiate one from another.

Those bugs can sometimes carry disease. Most biologists who travel to Peru, Ecuador, Costa Rica, Guatemala, and southern Mexico get pumped full of shots and then stop at the local "pharmacia" to fill up on antibiotics in case something gets past their vaccinations. The tropical jungle is beautiful but difficult.

With global warming, the bugs will be out in force. The mosquitoes will start swarming with the increase in warmth and humidity, as will the parasites and viruses within them. Scientists who've studied mosquitoes in Iquitos, Peru, find that a shift in climate of only 1.8 to 3.6°F (1 to 2°C) can have dramatic effects on the spread of mosquitoes and the diseases they carry. Mosquitoes that carry dengue and yellow fever have been reported at elevations of 7,220 feet (2,200 meters)—a lot higher than normal—in Colombia following periods of heavy rain. This increases the threat to people who live at high altitudes. Models project a global increase in the number of people at risk of malaria by 260 to 320 million due to an expansion of potential transmission zones. Increases in precipitation might drive cholera outbreaks as well.

The Polar Regions

Climate changes in the polar regions will be the greatest of any area on Earth and will cause major ecological, sociological, and economic impacts—especially in the Arctic and Antarctic Peninsula. The Arctic warmed as much as 9°F (5°C) in the twentieth century already, and sea ice has decreased by 2.9 percent per decade. The permafrost has begun to thaw in places, and this mucks up highways and slows traffic. In the Antarctic Peninsula, a marked warming trend has caused spectacular losses to some ice shelves.

Scientists from the British Antarctic Survey studying satellite evidence report that glaciers in the Antarctic Peninsula are flowing faster, further adding to sea level rise. As warmer air and ocean thins the glaciers, the weight of the ice reduces, allowing the glaciers to slide faster. The more that is lost, the less there is to hold back what's left.

There will be some benefits, such as a reduced demand for heating fuels. With a loss of sea ice, the Arctic Ocean may be open for more ships. And more ships will bring more commerce and more tourists.

The Winners and Losers

Indigenous people may be the least equipped to cope with changes in climate. Indigenous people number about 1.5 million of the total Arctic population of about 10 million people. They rely on the harvesting of fish, oil, minerals, and timber. These resources form the basis of their culture. The hunting of wildlife is an essential part of Inuit tradition. Hunting not only provides clothing and food but also contributes to community cohesion and self-esteem. Knowledge of wildlife and the environment strengthens community relations.

The sea ice is vital to a number of forms of hunting. If the ice doesn't form, hunting is delayed or doesn't take place at all. Walrus, whale, seal, fish, and birds are dried in the spring and summer to preserve them for the long winter months—but when air is too damp and wet, food becomes moldy and sour. The length of the wet season also affects the ability of indigenous people to gather foods such as willow leaves, beach greens, dock, and wild celery.

> **Global Warnings** _____
>
> The permafrost in the polar regions is an important substrate to permanent structures. In fact, a 3.5°F (2°C) rise in the soil temperature in the Yakutsk region of northern Russia decreases the bearing capacity of ground under buildings by 50 percent. If the permafrost melts, a lot of architects and engineers in the polar regions will have to go back to their drawing boards to rethink existing structural designs.

Technologically developed communities in the polar regions are likely to adapt a lot better. As the oceans start to thaw, increased commerce and tourism will support changes in roads, buildings, planes, and cars. It will be easier for oil companies to build offshore oil wells. They won't have to worry so much about the ice. However, as the permafrost melts, it will muck up the roads and undermine the buildings.

There will be new sea routes opening in the Northeast and Northwest passages. The fisheries for salmon, cod, shrimp, and herring may increase. Ships will be able to use these areas without strengthened hulls. There will be new opportunities for shipping oil, gas, minerals, timber, freight, and tourists. But they'll need new harbors, new international maritime laws, and more collision insurance to protect from inevitable spills.

Driving Others

The polar regions contain important feedback mechanisms that can increase the warming. Many of these mechanisms are self-amplifying—and once triggered, will affect other regions of the world for centuries to come:

◆ Snow and ice reflect heat very well, but if warming shrinks this white cover, the newly exposed ground will absorb the heat much more efficiently, which will further reduce the ice in a positive feedback cycle.

◆ Global warming could shut off the global ocean conveyor belt, the Thermohaline Circulation, with widespread and catastrophic effects.

◆ Projected warming will alter vast areas of oceans, wetlands, and permafrost, which act as major storage areas for CO_2 and methane.

◆ Reduced circulation in the Antarctic Ocean will decrease the ocean's ability to absorb excess carbon dioxide.

Small Island States

Some of the most concerned of the attendees at conferences studying or negotiating the issues of global warming are small island states. The Republic of Maldives—1,190 coral islets in the Indian Ocean southeast of Sri Lanka—has no tropical forests, hardly any automobiles, and little industry beyond the canning of bonito, yet its representatives are regular attendees at major global warming conferences. Their big concern is how much longer they will exist. Few of the islands have any point more than six feet above sea level.

Most of the island states of the world are located in the tropics or subtropics. They span the regions of the Pacific, Indian, and Atlantic oceans as well as the Caribbean and Mediterranean seas. Although it is hard to generalize about these varied communities, they share some common features. They are all surrounded by the oceans, have limited natural resources, large population densities, and poorly developed infrastructures. They also have limited funds, human resources, and skills. In general, they are sitting ducks for global warming and its attendant rise of cyclones and sea levels. And the effects would be dire indeed for coral reefs, water resources, agriculture, fishing, and the tourism industry:

Global Warnings

The 200,000 people living on the 1,190 islands of the Republic of Maldives may be some of the first to suffer from global warming and its predicted rise in sea levels. Few spots on this Indian Ocean archipelago, with its 2,000-year-old culture, rise more than six feet above sea level.

◆ Coral reefs, mangroves, and sea grass beds of coastal mainlands will be severely threatened by rising sea temperatures and sea levels. Changes in sea levels could drastically affect mangrove forests, which provide vital protection for tropical coasts, bays, and estuaries.

◆ Water resources and agriculture are critical to small island states. Changes in the height of the water table and an increase of salt in freshwater from rising sea levels could be catastrophic for many crops.

◆ Tourism is the major source of revenue for many island nations. Almost all settlements, structures, and tourist activities are located along the coasts. Changes in temperature and rainfall—as well as the loss of beaches—could devastate the economies of many small island nations.

As you can see from the examples in this chapter, the effects of global warming would be devastating to many regions on our planet.

The Least You Need to Know

◆ Global warming will gift Africa with floods, droughts, famine, and disease.

◆ Asia will have more trees in the north and a few more crops in the midlatitudes, but will lose land in the southeast delta regions.

◆ Australia and New Zealand will get hotter and drier and lose plant and animal species.

◆ European cities and beachfront resorts will be hot and muggy in the summer, and the Alps won't get as much snow in the winter.

◆ Latin America will get buggier in the tropics and stormier in southern Mexico and Central America.

◆ The polar regions will sustain some of the greatest changes, and some of those effects could radiate out into the rest of the world.

◆ Small island states might sink under the rising seas.

On the Home Front

In This Chapter

- ◆ North America gets toasty
- ◆ Floods but not enough water
- ◆ Trees on fire
- ◆ Lots more storms

Findings by the United Nation's Intergovernmental Panel on Climate Change (IPCC) as well as the U.S. Environmental Protection Agency (EPA) suggest that under global warming scenarios predicted for the next 100 years, North America will be quite a different place than it is today. Our water, forests, wild lands, coasts, houses, dams, and health will change as global warming begins to affect our climate. Let's take a look at some of the particulars.

North America vs. Global Warming

When the IPCC grouped countries in the western hemisphere, it referred to Canada and the United States as North America. They referred to Mexico, Central America, and South America as Latin America. Because this is one of the bases of its monumental study on global warming, I'll keep it that way for this book.

Like our Latin American friends, we're in for a hot time. Climate change will significantly affect natural ecosystems, where and how we farm, the availability of water to irrigate crops and produce power, the wetlands that spawn our fish, the beaches that we use for recreation, and all levels of government and industry. Our water resources are heavily managed, yet they weren't designed for what globally warmed weather might have in store for us. Insects and fire might play havoc with our forests. The rising seas will attack coastal areas. Storms will brew, and Florida might sink.

North America is a land of opportunity. Its gross domestic product is way up there. North Americans live long and spend lots of money. But they also use an enormous amount of natural resources and emit a lot of carbon dioxide. Although North America has a relatively clean environment, its use of fossil fuels for transportation and industry makes it the largest per capita producer of carbon dioxide in the world.

Global Warnings

Data from tree rings and lake levels in California show that in past millennia, there have been severe droughts of more than 80 years in duration. Evidence also suggests that there have been long periods when the Rocky Mountains and Great Plains regions were much drier than today. These incidents happened without the added push of global warming. And they caution us against assuming that tomorrow's climates will be as benign as today's.

North America appears unready for the consequences. It's complacent. North Americans view climate as a constant. Oh, it might vary a bit from time to time—a few storms here, a few droughts there—but nothing to worry about. But scientists who've studied the climate find that it is anything but constant. Even without the added risk of increasing the greenhouse gases, the climate has varied greatly in the past. Over the course of the twentieth century, there were periods of extreme weather. The drought of the 1930s altered large portions of American lands and society. In recent years, damage resulting from droughts, storms, flooding, hurricanes, and tornadoes has been on the rise.

Water

As the temperature warms, the ability of the atmosphere to hold more water will increase—and so will precipitation. But the picture won't be even. Climate change will affect different regions in various ways. The Southeast, Gulf, and Mid-Atlantic states might experience more floods, but in the West reduced snowfall and rising temperatures will increase the risk of drought. With less snow, there will be less snowmelt to feed important water sources like the Colorado and Columbia rivers. With higher temperatures and less snow, most of the flows could take place in winter and spring—leaving the summers, when snowmelt normally provides water reserves, much drier than normal.

Floods

The ability of the warmer air to hold more moisture should lead to an increase in floods in some areas. Subtropical conditions will extend farther north into the United States, which will change the precipitation, the vegetation, and the potential for disease in the southern states.

But with warmer temperatures come warmer rains and less snowfall. In mountainous areas of western North America, small, high lakes and snowbanks contribute to the bulk of the flow of many river systems. Warmer winter precipitation will not only increase winter and spring flows but will also increase the chance of rain on snowfall—a combination that provides some of the greatest risk of floods. In the Southwest, increased rains may also lead to increased flash floods—a result of the fact that the desert landscape has so little vegetation to hold back moisture that it can't handle rapid increases in runoff.

The water resources of North America have been heavily modified to increase the long-term ability of man to use the water. Private, state, federal, and provincial governments have built everything from private ponds to major dams. Water management practices like these have made it feasible for most North Americans to have access to water in years of both high and low flows. There has been a price to pay, however. Dams and levies often radically alter the natural ecosystems. In some areas, dams are being removed or considered for removal to restore native fisheries.

The human-made reservoir capacity in North America is equal to approximately 22 percent of the average annual runoff. That compares to an average of 10 percent worldwide. In the United States alone, there are 75,000 dams with a combined storage capacity equal to 70 percent of the average annual runoff. But storage capacity was built on past water records. Toss in a flood or a drought, and you might get a system that still won't meet increased demands.

In addition, levees and dams are often successful in managing most variations in the weather, but when precipitation gets heavy and the dam breaks, the damage can be a lot worse than if there had been no dam at all. What happened when Hurricane Katrina destroyed the levees in New Orleans in 2005 is a tragic example of those consequences. One community's efforts to protect itself from floods through the construction of dams or levees might just end up increasing the likelihood of flood damage to another community downstream.

Heavy rains can cause radical disturbances to the land and destroy freshwater and marine environments. The rainfall from Hurricanes Dennis and Floyd in eastern North Carolina in September 1999 provides an example. Rainfall of 3 feet (0.9 meters) breached containment ponds for poultry and hog waste and turned the flooding waters into a soup of raw sewage, fertilizers, and decaying vegetation. The floods contaminated the groundwaters and eventually reached the waters of the estuary between the Carolina mainland and the Outer Banks and turned the water brown. That area is one of the main nursery grounds for commercially important fish from Maine to Florida.

Droughts

Although some areas may get increased precipitation, global warming may subject the Midwest of the United States and the Canadian prairies, the agricultural heartland of the continent, to severe droughts. The natural environment might actually shift from grassland to something a lot closer to desert.

Heat not only reduces snowfall but increases evaporation. The latter may reduce the water levels of the Great Lakes and the outflow into the St. Lawrence River. Reduced snowpack and increased evaporation might also lead to reduced stream flows and lake levels in many parts of Canada—despite the fact that scientists predict increased annual precipitation. Declining snowpacks and increasing temperatures will essentially counteract the effect of more rain.

This is the problem in the West, too. Declining snowpacks will lead to heavier winter and spring rains and floods but decreased summer snowmelt, which will lead to decreased summer water. This will affect both wildlife and agriculture. In the arid Southwest, this could lead to increased reliance on groundwater and a lowering of the water table. The end result could be that more water is moved from agriculture and put into urban use.

Your Water Bill, Sir

Will all these water problems increase your bill? Is the pope a Catholic? Even in the face of hopeful increases in the conservation of water, the projected increases in water demands for the growing populations would cost $13.8 billion more in 2030 than in 1995. This is in terms of dams, aqueducts, and the purchase of water rights. And we're not talking global warming here. This all comes from the high costs of developing new water supplies, environmental concerns, a growing appreciation of natural streamflows, and efforts to improve water quality. Now add some global warming to the equation, and the projected annual increase will go from $13.8 billion to $105 billion—about $308 per person annually. A family of five will pay about $1,540 a year more for water than they do today.

Forests and Fires

North America contains about 17 percent of the world's forests. Temperature and precipitation strongly affect the type and density of forests. Climate change at predicted levels could eventually move the southern boundaries of most North American forests as much as 400 miles (600 to 700 kilometers) northward. This is because the southern limits of forests will become too hot for individual forest types, and there will be massive die-offs. The northern boundaries, however, would move only as fast as the normal rate of migration for forests. Projecting a migration rate of 60 miles (100 kilometers) per century—about double the known historic rate—the ranges of forests might shrink simply because the southern boundaries may advance more rapidly than the northern perimeters. Forests might climb higher in mountain environments, although at the expense of increasingly rarer arctic-alpine meadows and tundra. In general, the north *boreal forest* will advance poleward as the permafrost unfreezes.

Global warming should be easier on managed forests because humans can help the forest adapt to change. But it will be harder on native forests that don't have that added manmade assistance. If climate change results in the wholesale replacement of one forest type for another, wildlife habitat and natural ecosystems could be altered dramatically.

def•i•ni•tion

The **boreal forest** is the most northern type of forest and is dominated by coniferous trees such as the evergreen spruce, fir, pine, and the deciduous larch or tamarack.

Global warming might significantly alter forest types and reduce the land area of healthy forests. A study of forests in northern Mississippi and northern Georgia indicates that seedlings currently in such areas would not grow due to higher temperatures and drier soils that global warming would bring. In central Michigan, grasslands with some sparse oak trees might replace forests now dominated by sugar maple and oak. In northern Minnesota, the mixed boreal and northern hardwood forests could become entirely northern hardwoods. Changes could begin in only a few decades.

Drier soils expected to accompany global warming could lead to more forest pests and diseases. Add heat, dry soils, and some extra pollution to a forest and you have the perfect combination for a forest fire. A lot of this depends, however, on the amount of warming. If there is only a small amount, then the increased warmth and increased atmospheric carbon dioxide could actually promote forest growth. But as the warming continues to rise, increased water use associated with higher temperatures overwhelms the carbon dioxide effect, resulting in fewer trees. Plus, higher temperatures bring increased disease and insect problems.

Tourists Go Home

The United States and Canada are among the top 10 tourist destinations in the world with 1998 receipts of $71 billion and $9 billion, respectively. However, tourists are apt to visit a place that is quite different than the ones the postcards and the coffee-table books depict. Parks and other natural areas are important tourist destinations based to a considerable extent on the species they preserve and the natural environments they sustain.

One study revealed that 75 to 85 percent of Canada's national parks system would experience major shifts in the dominant vegetation under global warming scenarios. Large areas of the Yellowstone National Park region would experience extinction of a number of plants.

Hot Debates

Golfers may be some of the biggest losers in a globally warm world. Increased municipal and agricultural demands for water might inch out plans for developing new courses—and, in times of drought, diminish the capacity to irrigate existing facilities economically. Although the warmer weather may extend the recreational season, the deleterious effects that global warming might have on recreational lands will spoil the extended party.

Then there are beaches. Coastal zones are among the best tourist destinations in North America, accounting for up to 85 percent of all tourist dollars spent in the United States. But sea level rise will wreak havoc, particularly to beaches backed by seawalls or development that can't migrate backward. One study estimates that just to replace the sand on U.S. beaches that would be lost to a 20-inch (50-centimeter) sea level rise would cost $14 to $21 billion.

Global Warnings

And then there are the bugs. Higher temperatures and warmer winters could reduce the winterkill of insects as well as broaden insect ranges. Toss in a lot of rain, and that tends to reduce the efficacy of pesticide applications. Plus, it can move pesticides into areas where you don't want them.

Agriculture

In most areas of the United States, global warming will reduce yields of corn, wheat, and soybeans with individual farm losses ranging from small amounts to 80 percent of the crops planted. The decreases would be primarily the result of higher temperatures, which would shorten a crop's life cycle. In Canada and the northern portions of the United States, yields of corn and soybeans could increase as warmer temperatures extend the frost-free growing season. Southern areas growing heat-tolerant crops such as citrus fruits and cotton might also benefit from a reduction of killer frosts. Still, yields may decline in southern Florida and Texas because of excessive heat during the winter.

In general, global warming and increases of carbon dioxide in the atmosphere will work both ways, making some crops grow while others wither on the vine. Increased yields in the mid- and high latitudes may come as a result of the positive effects of carbon dioxide (remember, plants breathe that stuff), a longer growing season, and a lessening of cold temperatures. On the other hand, decreases in yields in drier areas will come courtesy of a lack of water.

An increase in storms won't help, either. One of the major reasons that the corn belt region of the United States is so productive is that temperatures don't vary that much from year to year across the region. The main risk of climate change to some regions of North America may be primarily from an increased variability. Corn, wheat, and soybeans know what they like, have grown accustomed to the weather on the local farm, and—like many of us—don't like a lot of change. Climate change, however, appears to be on the horizon.

Ranching

Global warming may increase livestock appetites while it alters grasslands and feed. Most estimates suggest that the negative effects of hotter weather in summer outweigh the positive effects of warmer winters. An expected increase in storms could breed havoc, though. The severe ice storms in eastern Canada and the Northeast United States in the winter of 1998 were brutal on livestock in those regions.

Fisheries

Variations in the climate play a big role in a number of North American coastal and deep-ocean fisheries. Changes in sea temperature, ocean water nutrients, and circulation patterns that are expected with most global warming scenarios may affect the populations of various species of fish. Climate variations may reduce the abundance of microscopic plant and animal life in the sea, and this can cascade through the chain of marine predator-prey relationships.

Managing fisheries is going to be a rough job in the presence of climate change as managers try to figure out where the fish are, how many they have, and who is entitled to what. Recent history is not on their side. The Pacific Salmon Treaty was embroiled when declining runs of southern coho and chinook salmon were coupled with the increased abundance of salmon in Alaska. That threw a monkey wrench into efforts to achieve a mutually acceptable balance of U.S. and Canadian interceptions of one another's salmon. The collapse of the cod stocks off Newfoundland on Canada's east coast is another example of the inability of government to control a fishery when coupled with less-favorable environmental conditions.

Human Health

Human illnesses and deaths are variously linked to weather patterns. Bad weather brings greater incidence of flu and pneumonia. It can also whip up the pollen, which has effects on asthma and allergies. Death rates increase for the sick and the elderly when extreme weather arrives. Increased temperatures and rain also affect disease-carrying mosquitoes and ticks. They bring Lyme disease, Rocky Mountain spotted fever, and dengue and yellow fevers. Global warming might also bring an increase in pollution, which can heighten the incidence of asthma, emphysema, and other respiratory diseases.

In the United States, populations in northeastern and midwestern cities may experience the greatest number of heat-related illnesses and deaths based on past experience. Heat waves in urban areas in these regions have been brutal. Heat killed 118 people in Philadelphia in 1993, 91 persons in Milwaukee in 1995, and 726 persons in Chicago in 1995.

Frosts and Heat Waves

In a warmer world, heat waves are expected to be more frequent and severe. And that can cause increases in illness and death, particularly among the young, the elderly, the poor, the ill, and those who live on the top floors of apartment buildings without access to air-conditioning. Still, more people die of exposure to cold than exposure to heat. So because there will be fewer cold waves, there should be a decrease in weather-related deaths.

Malaria, Anyone?

Malaria is today uncommon in the United States and Canada. But it wasn't always that way. As recently as 1890, there were more than 7,000 malaria deaths per 100,000 people in the United States and southern Canada. In Canada, the disease disappeared at the end of the nineteenth century. By 1930, malaria was under control in the northern and western United States and generally caused fewer than 25 deaths per 100,000 in the South. In fact, in 1970, the World Health Organization's Expert Advisory Panel on Malaria recommended that the United States be included in its register of places where malaria had been eradicated.

Malaria is still found in tropical and subtropical regions. But a number of studies indicate that with global warming, malaria may drift northward from its current tropical residence and be reintroduced into the United States and southern Canada. Public health measures will play a large part in the existence or spread of the disease in a globally warm future.

Storms

Have we got some storms for you! The U.S. Federal Emergency Management Agency (FEMA) declared fewer than 20 natural disasters annually in the 1950s and 1960s, but more than 40 a year in the 1990s. Global warming could up that ante. We've already talked about floods. There will be lots of them, especially in winter and spring. Sea level rise could increase the tidal surge damage that will come from hurricanes. A growing number of scientists now believe that an increase in ocean temperatures will drive an increase in the intensity and duration of hurricanes and perhaps even the overall number of these deadly events. Snowstorms might be down, but areas susceptible to freezing rains will increase. These severe winter storms are rough on power transmission, and the combination of severe winter storms and loss of power can have devastating consequences even in highly developed regions such as Ontario-Quebec and the northeastern United States.

In one such storm in January 1998, more than 31 inches (80 millimeters) of freezing rain—double the amount of any prior ice storm—produced the largest estimated insurance loss in the history of Canada. Power poles and power lines gave in under the weight of all that ice, stranding some residents and farmers without power for as much as four weeks. Almost 5 million people were without power at some point during that storm. Total Canadian losses were about $4 billion in U.S. currency. With global warming, ice storms will become more common.

Climatoids

An example of the type of storm manipulation that the world might become more used to in a globally warm environment is the cloud-seeding program in Alberta, Canada—underway since 1996 to reduce the severity of hailstorms. Central North America is one of the most hail-prone regions of the world, with hailstorms accounting for more than half of its catastrophic weather events. A 1995 hailstorm did more than $1.1 billion damage in Dallas; a 1990 storm did $600 million damage in Denver; and a 1991 hailstorm did about $240 million ($340 million Canadian) damage in Calgary.

Wetlands

About 14 percent of Canada's surface area is covered with wetlands (about 24 percent of the global total). About 6 percent of the United States is wetlands. Human activity has greatly affected midlatitude wetlands over the last 200 years. Agriculture and development have destroyed more than half of these vital areas in the United States. Global warming will affect what's left.

Sea level rise will alter coastal wetlands in many areas with potential damage to ocean fisheries. Increased drought conditions in the Prairie Pothole Region of the northern Great Plains will have drastic effects on U.S. duck populations.

You Wouldn't Want to Live by the Coast

Some 65 percent of people in North America live in coastal communities. This includes people who live near the Atlantic and Pacific oceans but also those who live near the Great Lakes. Everybody is going to see some changes. The worst will be for those on the coast who may receive an occasional combination punch of storm surge and sea level rise. As the seas rise, saltwater will work its way into estuaries and other aquatic systems with injurious effects to wildlife, agriculture, and industry. High seas and more floods will make it increasingly difficult for those who live in coastal bays and river deltas.

Coastal ecosystems are clearly vulnerable to changes associated with sea level rises and increased runoffs. The increased frequency in hazardous blooms of algae in estuaries in the Gulf, Atlantic, and Pacific states has threatened both the commercial fishing industry and public health. Consumption of toxic algae by shellfish can create concentrations of toxins in the shellfish that are high enough to cause paralysis and death to humans unlucky enough to eat them.

Got Insurance?

As shellfish poisonings are piled on top of ice storms, catastrophic floods, and storm surges, the tab is going to run up. Recent increases in financial losses from hurricanes, winter storms, and floods have raised questions about what it's going to be like when we toss in some global warming. The National Climatic Data Center reports that from 1980 to 2006, the United States has sustained 70 weather-related disasters in which overall damages and costs reached or exceeded $1 billion at the time of the event. Sixty-one of these disasters occurred since 1988, with total unadjusted damage costs of approximately $430 billion. FEMA reports the preliminary estimate of damages from Hurricane Katrina is approximately $125 billion, making it the most expensive natural disaster in U.S. history.

Florida Goes Down with the Ship

The prospect of sea level rises from 4 inches to 3 feet (.09 to .88 meters) over the next century is one of the most dangerous impacts of climate change. The greatest impact will likely be in recently developed coastal areas like Florida and parts of the U.S. Gulf and Atlantic coasts. Insured property value in Florida alone exceeds $1 trillion.

In Florida, sea level rises of only a few inches will, because of the gradual slope of the state's coastal areas, advance the shoreline greatly. In some cases, the horizontal advance will be 150 times the vertical rise.

Sea level rise may advance as much as 400 feet in low-lying areas. This will flood shoreline homes and hotels and erode the state's treasured beaches. Sea level rise might also damage the Everglades, offshore islands, and coral reefs. Global warming might also increase forest pests, which will make the trees ripe for fire. In short, much of what has made Florida so precious might be lost in a globally warmer world.

The Least You Need to Know

- Global warming will bring North America floods, droughts, and high water bills.

- Forests might benefit or shrink depending upon the amount of warming.

- Parks, beaches, and other tourist attractions will suffer as the things that draw the tourists are destroyed.

- With global warming, we might get deathly sick—but at least we won't freeze to death as often.

- Storms like Hurricane Katrina might become a regular part of the human experience.

- Sea level rises might take their toll on developed coastlines like Florida and other Gulf and Atlantic states.

Chapter 15

Nature Coughs Up Her Share

In This Chapter

- ◆ Polar bears get hungry
- ◆ Bighorn sheep get chased
- ◆ Songbirds get scarce
- ◆ The place is crawling with fire ants

If the climate did become warmer, animals could move north. They've done it in the past. When the great glaciers of the last Ice Age advanced south and later retreated north, the wildlife followed them back and forth. In fact, English naturalist Charles Darwin in the 1850s asserted that if "they all migrated in a body together, their mutual relations will not have been much disturbed."

But the transition from the ice age to the present age, the *Holocene*, took thousands of years—during which time the average global temperature increased 9 to 12.5°F (5 to 7°C) and the sea level rose 328 feet (100 meters). What we may do in the next 100 years is to ramp up the speed of that temperature change by about 50- to 100-fold. It's not that same gradual increase. Things won't be so smooth. Some transitions are occurring right now, and they don't look choreographed.

Mammals

Climate changes have had enormous effects on wildlife in the past. Scientists believe that the Younger Dryas, the big dip back into the Ice Age that happened at the beginning of the Holocene, might have been a factor in the extinction of the great ice-age mammals—the mammoth, mastodon, native camel, ground sloth, giant bison, and the saber-toothed tiger. The coming age of global warming could have as drastic an effect, causing wholesale extinctions across the globe. Some effects are already here.

Polar Bears and Ringed Seals

Every year around December, about 200 to 250 female polar bears give birth to twins or triplets in Canada's Wapusk National Park area. They are part of a population of about 1,200 polar bears in and around the park, some 600 miles north of Winnipeg in northern Manitoba. The females fast during their first months with their cubs, but as spring approaches, they grab their cubs and make a run for it out onto the ice. The females desperately need food and must reach the Hudson Bay before the ice breaks up and before ringed seals, the favorite food of the Wapusk National Park polar bears, disappear. It will take a family of polar bears about two to three weeks to make it to the ice, during which time each female will probably lose one or more of her cubs. It's a tough journey.

Once the family reaches the ice, it'll feast on seals. Mom will teach whatever number of cubs that survive how to hunt for themselves. If the cubs get through their first months outdoors, they'll spend the next two years by their mother's side, migrating to and from the Arctic ice packs hunting seals. Polar bears gorge themselves during the seven to eight months of the year when the ringed seal is available. During that time, they can triple their body weight. They'll need the extra weight to get them through the summer when the ice breaks up on the Hudson Bay and the polar bears move inland onto the summer tundra.

Today, ice melts in the Hudson Bay three weeks earlier in the spring than it did just 25 years ago. The increasing sun and early ice melt means there is less time for the polar bears to fatten up for the summer. Scientists have noted a 10 percent drop in the number of cubs born in the last 20 years. Adult bears are also 10 percent thinner. At a lower body weight, females have a tougher time nursing, which is part of the reason why there are fewer cubs.

But the problems for polar bears go beyond Northern Manitoba. Although the population still numbers about 22,000 to 25,000 animals, polar bears are in grave danger. The U.S. Fish and Wildlife Service recently proposed listing the polar bear as threatened under the Endangered Species Act. Polar bears are adapted for living and hunting on sea ice, but global warming is melting the Arctic sea ice at an alarming rate. According to the National Snow and Ice Data Center at the University of Colorado at Boulder, sea ice is declining at a rate of more than 23,000 square miles per year—nearly 9 percent per decade.

Polar bears are literally drowning from global warming, unable to swim the increasingly longer distances between land and receding sea ice. They depend on the ice to hunt seals—their primary food. With ice forming later in the fall and breaking up sooner in the spring, the time period bears can forage for food is shrinking throughout the far north.

The U.S. Geological Survey's Alaska Science Center discovered that only about 43 percent of the polar bear cubs in Alaska's Beaufort Sea are surviving their first year as a result of their shrinking ice habitats. Cub survival is down more than 20 percent from the late 1980s and early 1990s.

> **Global Warnings**
>
> As global warming has increased the length and intensity of the summer sun near the Hudson Bay, it's afflicting the skin of Inuit children with a condition many mothers have never seen before. It's called "sunburn."

Caribou

Global warming could affect caribou, also known as reindeer. Presently, the caribou are distributed over much of the higher latitudes of the northern hemisphere. Reindeer run more than any other deer. They protect themselves by gathering into large herds and shaking off predators through fast and tireless runs. When wild caribou get frightened, they follow the first animal that breaks away from the herd. Although it is usually the older, stronger animals that lead the herd, a frightened cub bolting for its mother will sometimes take the whole herd with it. The old-world reindeer and the North American caribou belong to a single species that has several forms and sizes. Males weigh anywhere from 216 to 605 pounds (125 to 275 kg), depending upon whether they live on the tundra (the skinny ones) or in the forest (the fatter ones).

During the summer, caribou wander continuously from one meadow to another in search of plants, fungus, and lichen upon which they feed. In some areas, their wanderings might take them 1,000 miles in one direction. As soon as the snow melts, they wander into colder areas, especially where the winds blow. The goal of these long migrations is to search not only for food but also for windy places where they can get away from biting insects: bot flies, horseflies, gnats, black flies, and mosquitoes. They suck a surprising amount of blood from the reindeer. The animals never really recover fully from these attacks until August, when the flies begin to die back. Global warming would increase the number of insects. It could also lead to the spread of anthrax and hoof-and-mouth disease. There is a chance, however, that it could increase food supplies in some areas if the caribou can stand the flies.

One point of contention between oil drillers and environmentalists has been the amount of disruption that is caused by oil pipelines to natural movements of caribou in the Arctic Circle. According to biologists, when caribou run across an unexpected obstacle on their route (such as a new gas pipeline or a frozen river), they can take as much as a month to figure out how to get around it. Meanwhile, they are subject to starvation and predation from the packs of wolves that follow them. Global warming is apt to increase this disruption by altering the terrain.

California Sea Lions Take a Dive

The California sea lion is another loser in a globally warmer world. This is one of the best-known seals because it is easily trained for the circus and is frequently kept in zoos. True California sea lions are found off the shores of southern California, the coast of Baja California, in the Sea of Cortez, and off the Mexican Coast. California sea lions are noisy animals. Their voices have a high-pitched bark that is very distinctive.

I accompanied biologists one summer to San Miguel Island off the coast of California—a summer rookery for the animal. There, they come to mate and bear their young. Females gather into harems and big bull males challenge each other in open-mouthed fights for access to the females. Males typically wear large gashes in their coats during the summer mating season. On San Miguel, there can be up to 20,000 sea lions and other seals during the California sea lion mating season, and the noise is amazing. It's like walking into a big stadium and somebody has just made a touchdown—only the cheering goes on 24 hours a day.

The diet of California sea lions consists chiefly of squid and some other species. The calves are born during the month of June. The mother returns to the water to fish soon after the pups are born to feed their hungry mouths. Sea lions prefer to hunt in shallower waters both by day and night. The waters off San Miguel are rich with food, and in good years about 75 percent of the pups make it.

But in the El Niño year of 1997, it was a different story. When the waters grew warmer, the normal prey species began to swim deeper and move farther north. The result was catastrophic. In non-El Niño years, pup mortality runs about 25 percent—but that summer it went up to 70 percent. With global warming presenting the possibility of a permanent El Niño, this catastrophe could become a regular occurrence.

Grizzly Bear and White Pine

The two last remaining grizzly bear populations in the lower 48 states reside in northwestern Montana and in and around Yellowstone National Park in Wyoming. In all, there are fewer than 1,000 grizzly bears left where 100,000 once roamed over the western portion of the United States and Mexico. Grizzly bears were brought under federal management when they were listed under the Endangered Species Act in 1975. At that time, fewer than 250 bears occupied the Yellowstone area. Since then, the coordinated efforts of state and federal agencies working with conservation organizations and private citizens have increased the population to more than 600 bears, and the Yellowstone population was taken off the list. Still, a globally warm world could put them back on that list.

The problem is, their favorite food is whitebark pine seeds. The seeds are much larger and more durable than the seeds of other cone-bearing trees, and it requires much less energy to get at them. Whitebark seeds are usually ripe around mid-August, at which time a host of birds, chipmunks, and other animals (including the grizzly bear) harvest the cones en masse. Bears at this time go after the whitebark pine seeds almost exclusively—especially adult females who must fatten up quickly if they are to produce cubs during the winter.

Climatoids

One of the grizzly bear's favorite methods of getting enough seeds is to let red squirrels do the harvesting for them. Red squirrels gather whitebark pine cones directly from the trees and store them in caches that grizzly bears love to raid.

The problem is that the whitebark pine is a sub-alpine species that exists in Wyoming at elevations from about 5,000 to 10,000 feet. As a warmer climate drives other species of trees higher, the whitebark pine could get squeezed out. Plus, the whitebark pine is subject to a number of diseases that could be affected by rising temperatures. Chief among them is blister rust. This fungus has already taken a toll throughout the range—as much as 90 percent in some areas. Mountain pine beetles also take their toll on whitebark pine trees.

Recent surveys show that blister rust infections in Yellowstone are still relatively low. Scientists estimate that only about 5 percent of the trees are infected. Biologists thought the area was unfavorable for blister rust, but global warming is making Yellowstone's environment more blister rust–friendly. Blister rust could dry up trees and make them more susceptible to fires.

The Sheep and the Lion

Desert bighorn sheep range over the Southwest from Texas to California and into Northern Mexico. There are about 20,000 of these animals in the United States today—a fraction of their former population. Bighorn sheep emerged from earlier evolutionary forms during the last ice age, when the glaciers stood at their southern-most boundaries. The sheep were then confined to an area of the once-lush south-western United States. As these ice walls retreated, vast expanses of grassland grew up in their wake. The bighorn passively followed the endless grasses and thrived.

As the Southwest became more arid, the desert bighorn adapted to an environment that sometimes reached 130 degrees. The desert bighorn can survive without water for months on end, getting most of their water by eating vegetation. They are normally secretive animals. It's hard to find one, but there are tricks. Biologists showed me a watering hole near Lake Mead on the Colorado River above Boulder Dam and told me to come back sometime when temperatures exceeded 110°F (43°C) three days in a row and I would see bighorn.

The bighorn have excellent vision—equivalent to a pair of binoculars—so I was told not to try to hide. Instead, I was to sit out in the open about 50 yards back from the water hole with a beach chair, an umbrella, and a cooler full of cold drinks and see what would happen. Sure enough, a group of more than 10 bighorn sheep came to the water hole a little after noon in the heat of the day. They hesitated for a while, but deciding the clown under the umbrella was obviously harmless, came in and drank and let me take pictures.

The bighorn like the high, rocky precipices where predators cannot follow. They have a unique double-shelled hoof that spreads over sharp, broken rocks and a soft, cushiony pad in the middle for traction on slick surfaces. The sheep are capable of nimbly bounding up and down incredibly sheer mountain cliffs, often dropping 20 feet between contact points. They can also follow two-inch-wide trails across steep cliffs.

Bighorn sheep are famous for their highly stylized fights between males. Two rams who haven't decided who's king will often feed within fighting distance—seemingly unaware of each other. Suddenly, both will rear up on their hind legs and begin churning rapidly toward one another. Bodies propel forward and down, adding gravity to the charge; heads flick forward at the last second and horns meet with a bellowing crack that can be heard more than a mile away. Often, the two gladiators achieve a combined speed of 45 mph (72 kph) in the charge. Evolution has given the bighorn a double-layered skull, cross-connected with bone and thick facial hide, which enables the animal to withstand these tremendous forces.

But desert bighorn are reluctant to leave their high, rocky refuges and move to flat lands between ranges where they can't outrun predators. Biologists literally have to pick them up and take them down the road to get them to spread into new territories. This makes them uniquely susceptible to climate change, because the animals that will survive best are the ones who can adapt and move.

However, bighorn susceptibility is more complicated than that. The bighorn are running into increasing problems with mountain lions. There are several populations in the Southwest where mountain lions have almost wiped out the desert bighorn.

In the early 1990s, when desert bighorn populations were plummeting in southern California mountains, California Fish and Game biologists radio-collared 100 animals to see whether they could pinpoint the problem. When the results were tabulated, they discovered that more than 70 percent of the bighorns that died in the study were killed by mountain lions. This is not the only instance where mountain lions have proved to be the nemesis of bighorn sheep. Mountain lion kills have reduced bighorns in the Sierra Nevada Mountains to fewer than 100 animals. In the San Andreas Mountains in New Mexico, hungry lions have eaten the herd down to a single female.

Eric Rominger, a professor at New Mexico State University who works with New Mexico bighorn, admits that the idea of the lions driving a herd into biological extinction goes against the classical predator-prey relationship in which the two populations are supposed to stay in equilibrium. In other words, as the prey dies back, so do the predators. But he claims that fire suppression policies have allowed forests to invade the normally bald rocky areas where bighorn are found. The trees then provide habitat for deer, which are not usually found in dessert bighorn habitat. This allows resident mountain lions to prey on deer and stay healthy although desert bighorn are declining. In an environment without the deer, the mountain lions would normally decrease as their prey decreased so that they would never be in a position of wiping out a prey animal. But in this special situation, humans have disrupted the environment and made "wipeout" a clear and present danger.

In a globally warm world, trees will ascend the mountains and create havoc for desert bighorn sheep as well as for a number of other animals. This shows how complex some of these natural biological relationships are and how human-induced warming could throw them out of whack with results that aren't expected.

Sea Birds

Adélie penguins are already feeling the effects of global warming in the Antarctic. Scientists studying the birds there find that only 1 out of 10 Adélie penguins survive the winter and return to nesting grounds to breed. They believe that a 5.5°F rise in sea temperature since the 1940s is the villain. The temperatures have caused the numbers of krill, an important food for the penguins, to plummet.

The 1997–1998 El Niño not only affected California sea lions but also proved devastating for California's endangered brown pelican. In the spring of 1998, researchers found only 280 nests in a colony in the Channel Islands National Monument off California, which normally contained 20,000. Climate change produced by El Niño had sent the brown pelican's normal fish prey deeper and farther north.

Sooty terns in the Dry Tortugas west of the Florida Keys are also experiencing problems with climate change. From 35,000 to 45,000 sooty terns return each year to Bush Key, a sandbar measuring 30 to 40 meters wide and 800 meters long, to nest and raise their young. But according to Fernando Colchero, a doctoral student at the Nicholas School of the Environment and Earth Sciences at Duke University, since the 1950s the birds have been arriving earlier and earlier. Where they once showed up in late April and early May, they now show up in January.

Colchero thinks it's the result of climate change either in the form of warming waters or changing currents that bring food to the terns. Says Colchero, "It is one of the most dramatic shifts in the nesting season for any bird species." The problem is this earlier breeding season makes the terns more susceptible to herring gulls, peregrine falcons, and other raptors that gather there in the winter and feed on terns and their young.

Waterfowl

More frequent and severe droughts could dry up many of the prairie potholes—small wetlands that provide breeding habitats for most of North America's waterfowl in the Great Plains and southern Canada. Models developed by scientists at Boston University suggest that within 50 years, climate change could wipe out 54 percent of the prairie potholes and about 60 percent of the duck population.

Songbirds Stop Singing

Climate change could cause major shifts in the range of many songbirds and some local extinctions. The forests of northern Minnesota and southern Ontario could lose up to 14 species of warblers. Warblers live off insects such as spruce budworms, which ravage local forests—and their absence could lead to devastating insect infestations.

The Kirtland's warbler in northern Michigan provides an example of how global warming effects on one natural system will challenge another. The bird is confined to a narrow area of jack pines that grow in sandy soil. Models of global warming's effects on the system predict that the jack pine is likely to migrate north with the warming. The trouble is, Kirtland's warblers are not likely to survive the transition. That's because the birds like to nest on the ground under young pines. But the soil to the north is not sandy enough to allow proper drainage for the nests. Without that drainage, water would accumulate and kill the nestlings. Scientists who study Kirtland's warbler predict that global warming could wipe the birds out within 30 to 60 years.

Global Warnings

The combined effects of the fragmentation of wildlife habitat and warming present one of the potentially most dangerous results from global warming. During the ice ages, many plants and animals survived by moving with the changing climate to another better-adapted area. But migration is not an easy deal in the twenty-first century because wildlife must migrate across freeways, farms, industrial parks, military bases, and cities.

No Breaks for Reptiles and Amphibians

Reptiles and amphibians have a rougher time of it than birds—mainly because they can't migrate with the weather as birds can. They are extremely susceptible to climate warming, which may be one of the reasons why amphibian numbers have already fallen around the world. Reptiles and amphibians are ectotherms, meaning that their body temperatures are not regulated by their metabolisms like you and me but instead by the sun and the outdoor temperature. Scientists studying the desert tortoise find that tortoises come out of their burrows earlier or later depending quite literally on the temperature. And they often have to warm themselves up on a hot rock before their externally regulated metabolism is warmed up enough to eat. When the temperature gets too cold in the winter, they go into hibernation and stop eating.

Will fly-fishing be another casualty of global warming? Good trout or salmon streams must be clear, free-flowing, and—most importantly—they must be *cold*. But global warming could change the chilly streams that sustain these prized fish. Roughly 10 million anglers spend about 10 days a year fishing for salmon and trout. The estimated value of these two kinds of fish is about $1.5 billion to $14 billion annually. Salmon are also an important part of the cultural heritage of northwestern Native Americans.

Why are trout and salmon so susceptible to global warming? That's because these fish thrive in streams that are 50 to 65°F (10 to 18°C). In many areas of the country, these fish are already living at the upper limits of their temperature range. That means that even small changes in temperature could render streams uninhabitable.

One study found that if we fail to do anything about global warming, habitat loss for salmon and trout could be as high as 17 percent by 2030, 34 percent by 2060, and 42 percent by 2090.

Insects

Butterflies such as the sachem skipper butterfly in the Pacific Northwest and the Bay Checkerspot butterfly in California are already responding to climate and weather changes. This could foretell changes in other species that occupy the same ecosystem as the insects.

We've already discussed how mosquitoes and ticks, which carry potentially deadly diseases, will increase their ranges into North America as the climate gets warmer and wetter. Red imported fire ants may also do the same.

If you live anywhere around Florida, North Carolina, Oklahoma, or Texas, then you probably know about fire ants. Fire ants are stinging insects that came here from Brazil—and, in many areas of the Southeast, have taken over ground from native ants. Fire ants make it impossible for many people in the Southeast to walk barefoot in their own backyards. Fire ants will attack farmers who've stopped to fix a plow or even livestock trying to get to feed. Global warming may help these ants spread to other parts of the country.

Coral Reefs

Global warming could wreak havoc on the world's coral reefs, especially those already under stress by pollution, physical destruction, and overfishing. If sea levels rise, that could submerge coral reefs too deep for their survival. If carbon dioxide concentrations increase in ocean water, that could inhibit the ability of coral to build up their protective skeletons. Warmer water also bleaches the coral by killing off the algae that gives the coral reefs their color. This bleaching makes coral weaker and more susceptible to disease, predators, and storms. Coral diseases are already on the rise around the world from warmer waters and increased levels of dust from droughts.

Tide Pools

Global warming could alter the habitat for Pacific Northwest tide pools, which are an important marine resource. If global warming creates more warm water in coastal intertidal zones, this could cause ochre sea stars—an important predator—to prey more heavily on mussels. The increase in predation on the mussels could upset the delicate balance of nature in this system and produce damaging side effects that could ripple out into the deeper ocean.

Our National Parks

The animals and plants in our national parks are in trouble if the climate warms and shifts vegetative zones northward. Plants can't pick up and move to a more hospitable environment if roads and farms are in the way. And animals can't move outside the parks if the area is developed.

If we let the remnants of wildlife populations remain in isolated reserves, they may become sitting ducks for local extinctions. One method currently being attempted to deal with this problem is to establish corridors between parks and other open lands. Corridors may provide an environmental safeguard should species need to move in the face of global warming. This might require wildlife bridges or tunnels to be constructed between open habitats as was done for the Arctic caribou when the Alaskan pipeline was built.

In southern California, biologists use a freeway underpass to connect the Chino Hills State Park and Cleveland National Forest. Corridors like these provide a necessary component to the natural balance of nature. Without the corridor, there is insufficient space for larger predators. Mountain lions won't breed, and the bobcats would inbreed and eventually die out. There would be an explosion of smaller predators like the fox, skunk, raccoon, and feral cat, which would then put increased pressure on small songbird populations, including the California gnat catcher and the Least Bell's Vireo—both on the Federal Endangered Species List. Corridors like this underpass would also allow wildlife the ability to respond if global warming creates serious challenges to their environment.

The Least You Need to Know

- ◆ Global warming is already affecting polar bears near the Hudson Bay by breaking up the ice earlier in the year and limiting the bears' access to seals.

- The grizzly bear might be affected if global warming seriously damages white pine in and around Yellowstone National Park.

- Parks, beaches, and other tourist attractions will suffer as the things that draw the tourists are destroyed.

- Global warming will impact sea birds, songbirds, and waterfowl. It will do a number on amphibians and reptiles, as well.

- National parks might soon have to be connected by freeway underpasses and overpasses for nature to flee the warming climate.

Chapter 16

What's in That Crystal Ball?

In This Chapter

- ◆ The weather 100 years from now
- ◆ How computer models do it
- ◆ What are the variables?
- ◆ How accurate are the predictions?

On a rainy day on the Edwards Plateau east of Austin, Texas, Robert Jackson—professor of biology and director of the Center on Global Change at Duke University—takes us down two rows of small plastic greenhouses to peer into the future of our climate. Jackson and United States Dairy Association (USDA) researchers Wayne Polley and Hyrum Johnson put together this experiment, composed of 60-meter rows of plastic-covered chambers erected over plots of grassland and fed air with controlled amounts of carbon dioxide. In the first row, the CO_2 concentrations range from 200 parts per million to 365 parts per million—the actual range from the last ice age up to the present day.

In the adjacent row, the plots of grasses are fed air with CO_2 concentrations of from 350 ppm to 550 ppm—the latter figure being what scientists fear the levels in the atmosphere may be by the end of the century. According to Jackson, "Based on fossil fuel emissions, the carbon dioxide concentration in the atmosphere should be going up twice as fast as it currently is.

However, natural systems such as the regrowth of the Eastern forests are currently taking up that CO_2, and we may be getting a free ride."

But according to Jackson, the free ride may be nearing an end. That's because at about 400 ppm, close to where we are now, the available nitrogen in the soil runs out and plants stop their increased growth—as well as their increased absorption of CO_2. The decrease in nitrogen constrains the plants' ability to take up the extra CO_2.

Considering the expected population increase, greater CO_2 use per capita, and the inability of natural systems to take up CO_2, Jackson says that "We may well be looking at increases per year that are double what they are now, with atmospheric concentrations as high as 800 parts per million in this century."

Robert Jackson with CO_2 Experiment.

(Photo courtesy of Michael Tennesen)

A Grand Experiment

As the inhabitants of this planet, we are engaged in a big experiment. Sometime before the end of the twenty-first century, we may have more than doubled the amount of carbon dioxide in the atmosphere from its preindustrial levels. We don't know exactly when it's going to happen, but it looks pretty inevitable. That means that the level of carbon dioxide, which was 280 ppm in the nineteenth century and is now 379 ppm, will rise to 560 ppm. It's the equivalent of dumping several hundred billion tons of carbon in the form of CO_2 into the atmosphere.

Even when we arrive at this landmark goal of 560 ppm, we'll have to wait a while to see the final results. It might take a few decades for the climate to come into equilibrium with this doubled carbon dioxide. It takes a while for the ocean to heat up. In all, it might take a century to fully appraise the damage—if there is any damage.

Climate Models

Right now, scientists are trying to predict what the answer to our dilemma may be. They don't all have such elegant experiments as Jackson. Many are using computer models to predict the future of climate change. In many cases, they're using some of the same computer programs that predict the weather. Only instead of running these programs a few days or even a few weeks, they've attempted to run the models for years, decades, and even centuries.

The models are quite fascinating. Scientists build these models to represent our real world, then they speed them up, turn the sun on and off, pump up the carbon dioxide, and see what happens. These are high-order computer simulations.

Cost vs. Benefits

Running these models isn't cheap. As Richard C. J. Somerville, professor of meteorology at the Scripps Institution of Oceanography, describes in his book *The Forgiving Air*, this is big science. He compares the science to astronomical telescopes, particle accelerators, and satellites. Indeed, much of the data that goes into making these long-range predictions is gathered from satellites.

The computers cost millions of dollars, which is a lot of money no matter how you look at it. And that doesn't include the building you put the computer in or the staff you use to run it. NASA funds most of it. But weather modeling also shows up as a line item on the ledgers of the National Oceanic and Atmospheric Administration (NOAA), the Department of Energy, and the National Science Foundation (NSF). NASA leads the way because space research yields much of the data for making these simulations.

Accurate weather reports—the foundation upon which global climate models are built—have tremendous value. Knowing the weather tells a farmer when it is best to plant and when it is best to harvest. But other industries such as shipping and tourism also benefit. Knowing what the weather will be like in the next several days gives these industries a competitive edge. Knowing what the climate will be in the next several years, decades, or centuries could help large industries and entire nations.

The First Weather Report

Lewis Fry Richardson, a mathematician and physicist who lived from 1881 to 1953, was the first person to try applying mathematics to the problem of weather forecasting. He did much of his groundbreaking work during World War I. Then a devout Quaker and pacifist, he resolved the conflict between his religious convictions and a desire to see the war up close by driving an ambulance in France. While going back and forth to the front lines to evacuate the wounded, he came up with the arithmetic to make the world's first scientific weather forecast.

Richardson had no supercomputer. In fact, he did all of his math with paper and pencil. He was driven by the idea that he could make the imperfect art of weather prediction into an exact science by reducing the evolution of the atmosphere into mathematical equations. He felt that if we could mathematically predict things like eclipses through calculations, then why couldn't we apply those same principles to the various aspects that make up weather and predict it as well?

Using calculus, he derived a number of mathematical equations to describe the physical laws that govern weather phenomena such as humidity, pressure, temperature, and wind. His idea was to divide the area of the forecast into a grid or chessboard and gather statistics from the center of each square to build a starting point. He could then factor that information into equations to determine how things would change. The statistics factored all at once for every square were the starting point for the prediction—what scientists called *initial conditions*.

def•i•ni•tion

Initial conditions is a term that refers to the starting point of a weather forecast that takes into consideration temperature, wind speed and direction, air pressure, and humidity measured from points in a grid overlaying the landscape.

Richardson's first forecast was highly inaccurate. He tried to predict the weather in six hours at a point near Munich, Germany, but was way off. Richardson suspected that his errors were due to poor observation of the initial conditions. In fact, his observing stations were neither numerous enough nor close enough. He also didn't take into consideration the three-dimensional aspect of the atmosphere. To accurately predict the weather, one has to know not only what the temperature, wind, humidity, and pressure is on the ground but what it is in the upper atmosphere. Richardson didn't have that option.

Not only were the initial conditions off, but the math was bad. Besides, it took too long. To figure out what the weather was going to be like six hours in advance took him several weeks. By then, obviously, the usefulness of the prediction had passed. But Richardson felt that he was on to something. He thought the problems could be worked out. Someday, somebody—somewhere—would solve the riddle. His book *Weather Prediction by Numerical Process* was published in 1922 and was the first modern treatise on dynamic meteorology. The book described his process in embarrassing detail. Although it was a rough start, it was a start. Richardson was on to something for sure.

The Evolution of Accuracy

Richardson's basic methods evolved. Observations both on the ground and in the air got a lot better. The math improved as well. The theories of how the various aspects of weather and climate react got a whole lot closer to the truth. Richardson figured that it would take a theater filled with thousands of people, each doing the arithmetic for one small part of the calculation, to get the job done in a reasonable amount of time. Computers made Richardson's idea of a theater full of people unnecessary.

In the late 1940s, a team of meteorologists and mathematicians using one of the earliest computers made the first successful mathematical weather prediction. By the 1950s, meteorologists were routinely making weather forecasts by using Richardson's basic methods. Today's climate models are based on those same ideas.

Data In

To predict the weather, we must again start out with today's initial conditions. To do that requires a vast amount of measuring instruments, scientists, communications, and a good deal of international cooperation. Measurements are taken at airports, weather stations, and aboard ships. Weather balloons are flown at various stations around the world twice each day to give us accurate readings at higher altitudes. Weather satellites give us a much better picture of cloud formations and the three-dimensional atmosphere.

Adding up the data.

(Adapted from Climate Change 2001: The Scientific Basis)

To predict the climate is something else altogether. We are looking for much broader values. We aren't asking whether it's going to rain tomorrow but rather whether it's going to be warmer in the next 10, 50, or 100 years. We aren't asking about the specifics of the weather picture but overall general aspects. Our grids are farther apart, and our information is not so detailed (the view is one of broad strokes only). We look less at which way the wind is blowing and at the temperature in Seattle but more at larger features like how brightly the sun will shine, how much carbon dioxide will be present, and how much ice there will be.

Hot Debates

Many ask, "If we can't predict the weather more than a few weeks out, how can we predict climate 100 years from now?" Scientists have had some skill predicting El Niños close to a year in advance, with generally improving results over time. El Niño is a climate phenomenon that produces storms in Peru, droughts in Australia, and effects all over the world. Because El Niños last a long time, affect a large area, and involve both the atmosphere and the ocean, they are not an aspect of weather but of climate. In fact, their prediction is our most successful effort at predicting climate to date.

Data Out

Weather services have gotten a lot better with their forecasts. Those from the richest nations utilize computers that equal more than a few theaters full of mathematicians. Not only do we have improved computers but we also have finer resolutions of the grid elements that Richardson used to make his first mathematical weather prediction. The finer the grid, the smaller the squares—and the more accurate the predictions. The typical resolution for today's weather prediction models is about 60 miles (100 kilometers) across each grid element.

Global climate models are much coarser. The grids are spread so far apart that a hurricane or a tornado could pass through unnoticed. Our normal, day-to-day weather can't be factored into these models. To run climate models out into the distant future brings most computers to their knees. And that's just doing the broad-stroke stuff. The little details, like, "Is it going to rain, snow, hail, or hurricane tomorrow?"—the stuff that's important to us on the ground—won't come out of those models. What comes out is the big picture—most notably, increased temperatures. Although it doesn't tell us about hurricanes, it does tell us that the tropical ocean that spawns hurricanes will be warmer. Because hurricanes run on heat, meteorologists extrapolate from that a possible increase in hurricane intensity. Because heat puts more moisture into the air, they also predict that those hurricanes will carry more rain.

Limitations

Weather forecasts have gotten better. We joke about the weather, but farmers, air traffic controllers, and ship captains know it's not like it was 30 or 40 years ago. Gross errors made in the 1950s and 1960s rarely occur these days. Meteorologists regularly predict weather as much as 10 days in advance. Back then, you were lucky to get one day. Meteorologists feel that one day we might be able to depict weather as much as three weeks in advance. Much beyond that, however, is unlikely. There is a limit to weather prediction. The atmosphere is not a cyclical, periodic phenomenon like the earth's orbit around the sun.

Global Warnings

When the Galveston Hurricane hit that Texas town on September 7, 1900, it was totally unexpected. The weather forecast in Galveston County's *The Daily News* that day was on page 2 and predicted rain with high northerly winds, clearing by late Sunday. But by late afternoon Saturday the deadliest hurricane ever to strike the U.S. mainland came ashore and took the lives of more than 6,000 people.

Tropical Depression 12, as Hurricane Katrina was initially called, formed over the southeastern Bahamas at 5 P.M. EDT on August 23, 2005. The system began to intensify and was upgraded to Tropical Storm Katrina on the morning of August 24. Katrina was upgraded to the fifth hurricane of the 2005 season on August 25, only two hours before it made landfall in Florida. The results were catastrophic but not as catastrophic as if there was no warning at all.

It's variable, erratic, and random. It's like a pinball machine: no matter how good you are at pinball, when the ball leaves the lever and goes up the track, it hits a peg and bounces around. A small difference in the way the ball hits that peg makes a big difference in how it is going to bounce around. It's the same with golf, tennis, or baseball. A small difference in how the club, racket, or bat hits the ball makes a big difference in where the ball ends up. Mathematicians call it "sensitive dependence on initial conditions."

The atmosphere is similar. It's extremely sensitive to initial conditions. Vary those initial conditions just a little bit and you can end up with a totally different weather outlook. The atmosphere is as unstable as a golf swing. The flap of a butterfly's wings can disturb the initial state, and you can end up with a totally different outlook. This was the theory of Edward Lorenz at the Massachusetts Institute of Technology (MIT). He postulated that if you had a planet like Earth—and it had a butterfly that flapped its wings just once—that flap would eventually result over time in a weather picture very different than the weather on Earth.

But when we look at the broader-based climate, there are indications that this sensitivity to initial conditions may not play as strong a role. Our ability to project an El Niño is an example of a successful attempt to predict long-range weather—the definition of climate. Still, using weather-predicting models to project long-range patterns in climate has an inherent weakness. Mainly, we haven't done it before. We're on new ground. We're out there making new science. It's untested, untried, and has major consequences.

Feedback Mechanisms

Meteorologists believe that in making long-range climate predictions, the crux of the problem is determining to what extent various feedback mechanisms will affect the climate. When global climate models are run on our experiment of a doubling of the CO_2, the models predict an expected worldwide increase in temperature of 1.8 to 7.2°F (1–4°C). The variation is primarily due to how the models handle feedback mechanisms such as ice, soil, ocean, water vapor, and clouds.

The difficult part of climate modeling is predicting how those feedback mechanisms will work. If there weren't any feedback mechanisms, we could figure out the temperature increase on Earth from a doubling of carbon dioxide by simple math. The answer is that the world would increase its temperature by about 1.8°F (1°C). The models, however, show that the Arctic temperatures could rise 21.6°F (12°C)—and that's because tremendous feedback mechanisms come into play in the Arctic.

Ice

According to global climate models, if you double the amount of carbon dioxide in the atmosphere, the greatest amount of warming will be in the far north. The reason is because in the Arctic, there are vast expanses of ice and snow. And as the atmosphere warms up, the ice will start to melt. Already, 40 percent of the Arctic ice has melted. As the temperature gets warmer, the melting will accelerate. By the time the world has reached equilibrium from the doubling of carbon dioxide, half, three quarters, or even all of the ice will have melted in some areas.

Darker oceans, bare land, and grass-covered land will replace the ice. These are all darker than ice or snow and have lower albedos, or lower reflectivity. So instead of the sun's radiation being reflected back, it will be absorbed by the darker surfaces. Anyone who's walked barefoot to the beach in the summer knows that the darker asphalt road is a lot hotter to tread upon than the lighter concrete sidewalk. That's because the dark asphalt is absorbing heat that the lighter concrete reflects. In the process, the concrete stays cooler relative to the asphalt.

Computers predict that the greatest warming will be in the north because there is a positive feedback system. As the ice melts, the darker water or earth beneath it absorbs more heat. The change in the ice due to warming amplifies the warming. Estimating how much of an effect this will have accounts for some of the variation in climate-model predictions.

Climatoids

The amplification of the global warming that computer models predict for the northern latitudes is caused by the ice-albedo effect in which melting ice leaves surfaces that are more heat-absorbent. The effect is more pronounced in the high northern latitudes than in the high southern latitudes because the southern hemisphere has less land to be covered with snow and ice.

Soil Moisture

And then we have the tropics, where there isn't any ice. But there are feedback mechanisms there as well. One of the problems modelers have in the tropics is determining the amount of feedback in soil moisture. The reason is because when the sun's rays hit the surface of the earth in a place where the soil is moist, part of the energy of the incoming rays is used to evaporate the water. If the soil is dry, then all the energy in the earth's rays goes to heat the soil, rather than evaporate the water. If the sun has to tackle soil moisture, that's going to take some of its heat away. If the soil is dry, then all its energy goes into the air—making it hotter. This is also one of the reasons why the surface temperatures in the desert can get hotter than the tropics, although the tropics are closer to the equator and thus receive more direct light from the sun. Desert soils are dry and absorb more heat.

Rain and Snow

Scientists estimate that if nothing is done about the rising levels of greenhouse gases, then the oceans will rise about 20 inches (50 centimeters) by the year 2100. The reason is because as the ocean becomes warmer, it expands—warmer water takes up more room than colder water—and the ice that is on the land melts and flows into the sea. In short, warming causes both expansion of water and melting of ice, and these both lead to sea level rise. During the ice age, when more of the water was contained in glaciers, sea level was much lower than it is now. There was a land bridge that connected Alaska and Siberia.

But one of the things that happens with global warming is that as the temperatures rise and the ocean warms, more of the ocean evaporates into the atmosphere—and it rains and snows more. As a result, Antarctica and Greenland will lose ice into the sea because of increased temperatures and increased melting—but there will be more snow, so the amount of polar ice could actually increase.

Water Vapor

Climate models have long included a positive feedback mechanism from water vapor. Warm air holds more water than cold air. Air in the tropics is more humid than air at higher latitudes. As the temperature warms up, so will the ocean. A warmer ocean will evaporate more water into the atmosphere. And this additional water vapor increases the greenhouse effect because water itself is a greenhouse gas—more powerful than carbon dioxide.

In many ways, carbon dioxide is simply the trigger by which more water vapor enters the atmosphere. This water vapor will amplify the warming caused by the carbon dioxide. Because of this, the temperature rise from an increase in carbon dioxide might be 1.5 to 3 times as large as it would be without this feedback. Climate models have to deal with all of these interrelated phenomena in order to predict the future.

Clouds

Clouds may be the trickiest feedback mechanism of all. They have a dual role. First, they reflect sunlight coming in from the sun and redirect it back into space. Clouds reflect about 20 percent of the sunlight directed at Earth back into space. The effects vary with cloud height and type. Because less sunlight falls on the earth, less heat is absorbed, so it's cooler. Thus, clouds—by their reflectivity—cool the earth.

But, wouldn't you know it, they also warm it up. The clouds absorb the rays of the sun that bounce off the surface of the earth, and they redirect that heat back toward Earth (which increases the temperature). It's the greenhouse effect again. Remember, the glass over the greenhouse kept the plants beneath it warmer and moister. Well, that's just what carbon dioxide, methane, water vapor, *and* clouds do. They keep the earth beneath it warmer and moister.

So … which dominates? Satellites have been deployed to determine the contribution to the greenhouse effect and to Earth reflectivity from clouds. According to Richard Somerville, who has spent a good deal of his meteorological career studying the feedback effect of clouds, clouds (on average) cool the planet. In other words, the reflectivity of clouds dominates over their greenhouse effect. He translates the measure of their strength into watts—the same watts that power your light bulbs. In general, clouds deflect an average of 45 watts of sunlight per square meter of the earth's service. The greenhouse effect of clouds contributes about 30 watts of heat per square meter.

But that's without global warming. When you add the warming, cloud cover might decrease. Clouds currently cover approximately 50 percent of the surface of the earth. In a modeled double–carbon dioxide atmosphere, cloud cover is decreased by a few percent. Because the cooling effect of clouds dominates the warming effect, when you remove clouds you are removing more cooling than warming. The net effect is that it might get warmer. It's similar to removing the ice. The surface beneath is darker and absorbs more heat. Additionally, in a carbon dioxide–doubled world, the clouds on average are higher. And the higher the clouds go, the colder they get. Colder clouds don't radiate as much heat. According to Somerville, the difference in the degree of warming predicted by the models is principally due to how the models handle the difficult question of cloud feedback.

> **Climatoids**
>
> The earth reflects about 30 percent of the sunlight that lands on its surface. Two thirds of that is due to the reflectivity of clouds and one third is due to the reflectivity of ice, snow, and the bright surfaces of the desert.

The Forecast

All the climate models predict that the earth is going to get warmer. By the time the carbon dioxide rises to 560 ppm—double its preindustrial levels—the atmosphere could warm approximately 1.8 to 7.2°F (1 to 4°C). The IPCC predicts even greater warming by 2100, when CO_2 could rise from 540 to 970 ppm. Although there is a range of degrees, no one is expecting it to cool down. Taking the temperature of the earth is a little like taking the temperature of a sick patient. Is a 1°F (.6°C) rise in temperature enough to worry about? Maybe … maybe not. But what about 2°F, 3°F, or 4°F (1.2°C, 1.8°C, or 2.4°C)? That could indicate a serious underlying infection. And what about a 7.2°F (4°C) rise in temperature? That could indicate a fatal illness.

Should We Trust Them?

The trouble is, we've never done this before. Still, the best efforts of about 2,500 of the world's leading climatologists and meteorologists shouldn't be taken for granted—although even they have doubts themselves. But in 50 to 100 years, if their predictions prove wrong, we won't have lost much. We might even have a little gas left. We might have some new technologies that don't pollute the air as much. There may be some rainforests left. But in 50 years, if we've done nothing to turn the tide of increased greenhouse gases, it might be too late. That's because at whatever point we decided to stop pouring carbon dioxide into the atmosphere, the results we're looking at then might just have to be lived with for the next 200 years—about the time that carbon dioxide persists in the air.

The Least You Need to Know

♦ Increased growth in forests and grasslands may stop consuming our excess CO_2.

♦ Global-climate models run on the same machines that predict the weather. The difference is that weather prediction models try to predict detail, while climate models look for big-ticket items only.

♦ Weather models and climate models have lots of problems, but the models are still very useful.

♦ Snow, ice, water vapor, cloud cover, and other feedback mechanisms affect climate-model predictions.

♦ All models predict warming, but the range of temperature increase is broad.

Part 5

The Politics

Here we'll look at the Kyoto treaty. What is it all about? How was it developed, what does it require, and where does it stand now? How did this credit system work—trading smokestacks for rainforests? How does U.S. politics fit into this? What about the international debate—who's skeptical and who's not? Could pollution reduction, energy conservation, and rain forest preservation be worthwhile endeavors all by themselves?

In the late 1980s, the nations of the world came together to tackle the problem of ozone depletion. Certain gases, chlorofluorocarbons (CFCs), were creating a hole in the ozone layer over Antarctica. The national community gradually outlawed CFCs. It was a great moment. We had all gotten together and done something for the atmosphere. Why can't we do the same with global warming?

Finally, I'll sum up the argument for you. Who's behind the predictions, what are they saying, and what's the current evidence? Should we move forward, or are the costs too expensive?

"I give up. How many politicians does it take to screw in a fluorescent lightbulb?"

Chapter **17**

A Treaty Known as Kyoto

In This Chapter

- ◆ Bartering with carbon dioxide
- ◆ Cut pollution or grow trees
- ◆ Developing countries let off the hook
- ◆ The United States backs out, but the momentum continues

In late 2004, when Russia became the 126th country to ratify the Kyoto Treaty, it cleared the way for the international treaty on climate change to go into force. Multinational corporations that did business in signatory countries would have to comply with the new requirements. The pressure was on. Still, the United States and Australia never signed on to the pact— and China and India, although they signed, were not required to reduce emissions.

To many who had worked so hard for its passage, the failure to get some of the most polluting countries involved was a letdown. But the momentum toward establishing an international agreement to tackle this important issue has continued. Even in the United States, a number of cities and states have passed laws to bring their communities into compliance with Kyoto principles.

Perhaps it's a good idea to see how it all got started.

A Start

In December 1997, more than 160 nations met at the United Nations Conference on Climate Change in Kyoto, Japan, to develop an international treaty to deal with the growing body of evidence that global warming was a reality and that the world had to do something about it. At the Earth Summit in Rio in 1992, the Parties to the Framework Convention on Climate Change were given the task of coming up with an agreement that would halt the increasing flow of manmade greenhouse gases into the atmosphere. Five years later, the parties had come together in Kyoto to try to work out an agreement. It wasn't an easy task.

After 10 days of marathon meetings and late-night negotiations, an agreement was reached—and they finally put pen to paper. In the end, after 160 nations had wrangled, fought, and played tug-of-war, they passed an important milestone by acknowledging a global threat and coming up with a global solution to protect the earth's environment.

The problems it tackled were varied. We'd created carbon dioxide with automobiles, industry, cooking fires, plows, axes—even our breath. We had to take a hard look at the cost of modern transportation. What price had the atmosphere paid for the increasing efficiency with which we moved our products to the market? What were the real costs of cutting down the rainforests in terms of climate change?

It seemed now that the world might finally be ready to do something about it. Although the scope of the agreement was limited, its mechanisms were important because it set standards for further agreements and created tactics with which the battle could be fought and hopefully won. It recognized that concentrations of certain greenhouse gases had been growing since the beginning of the Industrial Revolution. And it recognized the international importance of reversing the steady increases.

Hot Debates

History played a vital part in the Kyoto negotiations. One of the reasons why industrialized nations were asked to go first in the negotiations was because of the longevity of carbon dioxide in the atmosphere. Much of the carbon dioxide that was released in the Victorian era by those nations was still in the air. Developing nations had only just begun to add to the problem.

Divvying Up the Burden

The idea was that 38 nations needed to reduce their greenhouse gas emissions by an average of 5.2 percent below 1990 levels. In reality, the reductions were parceled out according to who emitted the most gases and who could afford the changeover. The United States would reduce emissions by 7 percent below 1990 levels, Japan by 6 percent, and the European Union by 8 percent. The Russian Federation, the Ukraine, and New Zealand needed to return to 1990 levels while Australia and Iceland had to stabilize their releases at 8 percent and 10 percent, respectively, above 1990 levels.

The European Union wanted to get started right away, but the United States pressed to have enforcement of the reductions put off until 2008. This would provide a full decade to allow governments and industries to gradually shift emphasis toward increasing energy efficiency, to upgrade pollution equipment as older machinery wore out, to switch from dirty fuels to cleaner-burning energy, and to experiment with solar, wind, and other alternative forms of energy.

Getting 160 nations to agree on something wasn't an easy task. European Union delegates had arrived with the ambitious proposal for all industrial nations to cut their emissions by 15 percent below 1990 levels. It seemed easy for them to talk. The collapse of the former East German industries and a massive switchover in Great Britain from coal to natural gas during the 1990s had given Europe a head start on the rest of the world. But Europe relied heavily on emission-free nuclear power, and nuclear power wasn't exactly squeaky clean.

But the real struggle was in the details. Japan insisted it wouldn't go any higher than 5 percent in emissions reductions. A 2 A.M. phone call from then U.S. Vice President Al Gore to Japanese Prime Minister Ryutaro Hashimoto got Japan to up its ante. Gore argued that Japan couldn't wreck the chances to come to an agreement over a measly 1 percent. The Japanese prime minister relented.

There were other difficult points. At Kyoto, Russia and the Ukraine were given the right to emit as much greenhouse gas in 2010 as they did in 1990, the year after the Berlin Wall came down. But their industries had collapsed and their emissions were as much as 30 percent below 1990 levels already. Also, while the new treaty would impose binding limits on emissions by 39 nations, 121 less economically developed nations escaped any kind of reduction—even though some of them were expected to grow rapidly. It totally let both China and India off the hook. This was in spite of the fact that their emissions were growing rapidly and were eventually expected to lead the world.

The agreement placed limits on six greenhouse gases: carbon dioxide, methane, nitrous oxide, hydrofluorocarbons, perfluorocarbons, and sulfur hexafluoride. The European Union had wanted to limit the new curbs to just carbon dioxide, methane, and nitrous oxide. But the United States succeeded in adding limits on hydrocarbons, perfluorocarbons, and sulfur hexafluoride. The latter chemicals were arguably easier and cheaper to cut.

Is It Enough?

Would the Kyoto Protocol solve the problem of global warming? Hardly. According to the IPCC (the convention's scientific advisers), the treaty—if passed—would buy 10 more years at most. To halt global warming, they said, would take a much more radical commitment. The IPCC suggested that the world should agree to a maximum concentration of greenhouse gases in the atmosphere. A reasonable goal might be a doubling of carbon dioxide in the atmosphere over preindustrial levels. That figure represented an increase of about 50 percent higher than present-day levels.

Many had already argued that a doubling was dangerous. But scientists were ready to acknowledge that the doubling was likely to take place, treaty or no. But at least if it didn't get any worse, that would be something. Although the Kyoto agreements did not call for the levels of emission cuts that most scientists considered necessary, it was a start. It was a first step in the right direction—an acknowledgement of the problem and a resolve to find an answer.

Bartering

Despite its varied drawbacks, the Kyoto agreement did contain one breakthrough that many felt was significant. It allowed industrial nations to reduce emissions through a credit-trading system. The concept had been pioneered by former U.S. President George H. W. Bush in the Clean Air Act, which in the early 1990s cut acid rain in half by allowing U.S. utilities to trade sulfur dioxide credits. A polluting industry could buy or trade credits with another industry that had exceeded its pollution-control goals.

A similar idea was proposed at the Kyoto conference. The new system would permit the industrialized nations to trade carbon-emission credits among one another and with developing countries to reach their own greenhouse gas emission goals. Because global warming was an international problem, setting up a global trading regime seemed like a good idea.

A country with the potential for wind power or another source of renewable energy might develop them on a large scale and potentially reduce their greenhouse gas emissions to below required levels. They could sell the credits for the extra reductions to countries that didn't have as great a potential for renewables. It might be more cost-effective for an industrialized nation—particularly one who had already gone through significant pollution reduction—to simply buy the credits of another nation rather than to go through more expensive reduction methods.

The fear was, however, that some countries might find themselves with spare credits to sell—just because their economies were in the dumps. Such was the case with Russia and the Ukraine. The Kyoto agreement gave them pollution limits that were quite a bit above what they were producing in their then current economic downturn. Although Russia promised to invest any of the monies it might gain in greenhouse gas credit sales into cleaner technologies, some people saw this type of trading as a loophole in the basic role of global emissions reductions. Environmentalists referred to it as trading for *hot air*, a reference to the belief that this type of trade would yield little in the way of the conference goal of making concrete efforts to reduce greenhouse gas emissions.

def•i•ni•tion

Hot air is a term used by environmentalists to describe greenhouse gas reductions that resulted from weakened economies and the resultant reduction in manufacturing, rather than actual efforts put into making those industries operate cleaner and more efficiently.

Cleaning Up Someone Else's Backyard

In some cases, it was cheaper and easier to clean up someone else's mess rather than attend to stuff at home that has a higher price tag. Kyoto allowed for this. Under the agreement, a country could invest or participate in other ways in an emissions reduction program in another industrialized nation and gain credits that could be applied toward the emission reduction goals at home.

The treaty also allowed wealthy nations to invest in emissions reduction projects in poorer nations and get credit as well. Not only might it be cheaper to invest in emissions reduction in the third world, but that investment could become a major engine for getting clean-air technologies into poorer nations. This could prevent those nations from going down the same polluted path that industrialized nations previously navigated and now have to amend.

One of the basic beliefs behind the global emissions trading under the Kyoto agreement was that it didn't matter where one chose to lower emissions—whether at home or abroad. A reduction of the emissions on one side of the globe will eventually affect the other side of the globe in equal portions, principally due to the fact that the gases are so long-lived and eventually mix in equal portions over the entire world.

What About Methane?

The emphasis in the Kyoto agreement was on carbon dioxide. Of all the greenhouse gases, it is the longest-lived—lasting up to 200 years in the atmosphere. It is also responsible for the largest portion of the greenhouse effect.

Climatoids

Methane is the second-most important greenhouse gas, yet it is much more potent. Releasing 22 pounds (10 kilograms) of methane into the air today will warm the world about as much over the next decade as a ton (907 kilograms) of carbon dioxide.

Some scientists, however, felt that more attention should be given to methane. Methane has a much stronger greenhouse effect than carbon dioxide. The reason why carbon dioxide was given greater importance in the agreement was that methane only lasts a decade in the atmosphere. But a little bit of methane was equal to a whole lot of carbon dioxide in terms of climate change, so by decreasing methane there was a potential for a more immediate reversal of global-warming trends.

Plus, methane is easier to clean up in the atmosphere. All you have to do is fix the natural gas pipes so they don't leak so much. Put a little more soil or inert waste on landfills to encourage the growth of methane-eating bacteria. Maybe feed the cows something that doesn't give them so much gas.

In addition, cost-effective methods—such as plugging leaks in gas lines—would lead to savings over time, a counterpoint to those who argue that reducing greenhouse gases would hurt their economies.

Two Smokestacks for One Rain Forest

One of the methods for gaining emissions credits envisioned by the Kyoto agreement was to encourage the land to soak up more carbon dioxide. That meant creating man-made *carbon sinks* by planting trees or other vegetation or by practicing various types of soil conservation.

The treaty allowed industrial nations to receive some emissions credit for any land it had forested since 1990. However, they also had to increase their emissions reductions if they had deforested any of their land since that date. So, one could receive credit for creating a carbon sink but also get penalized for destroying a carbon sink.

Countries could also claim credit for planting forests in developing nations. The draw here was that an industrialized nation might plant a forest in the tropics, for example, where trees would grow a lot faster and land wouldn't be as costly. Not only would this decrease the amount of greenhouse gases in the air, but it would also decrease tropical deforestation (which had other benefits, such as increasing habitat for wildlife).

def•i•ni•tion

Carbon sinks are areas like forests, grasslands, and the oceans that naturally take up carbon dioxide from the atmosphere.

Shallow Plow

Countries could also claim credit for soil conservation measures, which would allow more carbon to be soaked up by the soil. One idea was *low-till farming*, a method of turning the ground that could trap more carbon in the soil. Deep plowing allows more oxygen into the soil and speeds up the decay of organic matter and the release of carbon dioxide into the atmosphere. Set the plow a little shallower, and it would decrease those emissions. Countries like Canada and Australia professed that by such soil conservation methods, they could soak up tens of millions of tons of carbon each year.

The European Union came into these talks feeling that these types of mitigation would only be a small part of the efforts to meet emissions targets—whereas the United States, Russia, Canada, and some other countries thought they should be a major part of the treaty. The traders eventually won out in the drafting of the agreement.

The Price of Carbon Dioxide

Although bartering seemed like a good idea, problems arose from trying to figure out how much to charge for emissions. In other words, how much is a pound or a ton of carbon dioxide worth? How does one accurately determine the value of a carbon sink? How does the world patrol individual country emissions? Will there be an international greenhouse police, or will each country monitor itself? Also, how will emissions trades be made? Will we print emissions credits? What bank will handle the transactions?

The problems of organizing and maintaining such an international monitoring and trade organization were difficult to say the least, but countries like the United States had success with trading sulfur dioxide emissions. Smaller municipalities like Los Angeles had allowed businesses to achieve credits for pollution reduction—not by adding new technologies to their smokestacks but by buying older, more polluting cars off the streets and sending them to the junkyard.

Your Bill, Sir

What all this would cost the economy was a point of contention. When Kathleen McGinty, chair of the Council on Environmental Quality for the Clinton administration, went before the U.S. Congress to report on the cost of the proposed treaty, she estimated that the cost of the treaty to the average family's energy bill would be about $70 to $110 a year in 2010. Industry estimates have been much higher.

Furthermore, McGinty noted that economists had estimated that the environmental, health, and economic costs of global warming projected to occur from a doubling of carbon dioxide in the atmosphere were about $80 billion a year. One of the secondary benefits of greenhouse gas reductions was a decrease in pollution. McGinty estimated that this reduction in pollution alone could result in a savings in health care costs equivalent to one fourth of what the United States would pay to achieve its greenhouse emission-reduction target.

Those in the U.S. petroleum industry were not the only ones concerned with the costs of implementing the Kyoto agreement. Although the Canadian government had pressed toward ratification of the treaty, Canadian industries—particularly in Ottawa—resisted. One report by the Canadian Manufacturers and Exporters Association claimed that 450,000 jobs would be lost because of Kyoto. The government countered that only about 61,000 would go—or, under a worst-case scenario, as many as 244,000.

Developing Countries

Under the Kyoto treaty, the world's developing nations were not required to make reductions in greenhouse gases—although Argentina and Kazakhstan nevertheless agreed to do so. The reasoning behind this was that industrialized nations, the United States in particular, were responsible for a disproportionate amount of greenhouse gases. The United States—although it had only 4 percent of the world's population—was responsible for 25 percent of the world's greenhouse emissions.

Because most of the carbon dioxide in the atmosphere to date was the responsibility of the industrialized nations, it was agreed that they would take the first step. Then, perhaps the developing countries would be more willing and eventually more able to follow suit. Poor countries insisted that the excesses of the rich nations were largely responsible for the problem and that cutting emissions would come at the price of economic growth—making their lives even poorer.

Yet it is the developing countries that will be producing the lion's share of global greenhouse emissions by the end of the twenty-first century. Argentina, at a follow-up conference on global warming in Buenos Aires, attempted to break the impasse by offering to make voluntary cuts and encouraging others to do the same. Countries such as China and India refused to debate the matter, yet China was expected to lead the world in the production of greenhouse gases by the end of the twenty-first century.

The U.S. Balks

Many in the U.S. Congress couldn't handle the lack of participation by the developing nations. A resolution in the Senate in 1997, which passed 95–0, put the Senate starkly against ratifying a treaty that required no meaningful participation by developing nations. The United States demanded that developing nations should accept their own specific emissions targets—even if those targets allowed emissions to increase.

But the biggest setback for the treaty came in 2001 when the Bush administration announced that it was explicitly opposed to the Kyoto agreement. Environmentalists from around the world greeted the announcement with loud dismay. Despite the growing evidence of global warming and the undeniable role of the United States in creating it, the administration had passed on the challenge.

Climatoids

Despite the fact that the United States has pulled out of the Kyoto agreement, many businesses have come to recognize global warming as a threat and the goal of reducing emissions as being in their economic interests. Several industries such as BP-Amoco, IBM, and Johnson & Johnson have voluntarily adopted emission-reduction goals that are stronger than those required by Kyoto.

The Rest of the World Forges Ahead

Some thought that when the United States backed out of the Kyoto agreement, this would spell death to the treaty. But in some ways, it has proved just the opposite. Japan and the 15 countries of the European Union have ratified the agreement.

Canada—although it had problems with the agreement—ratified it in December 2002. It seems that by pulling out of the agreement, the United States might actually have helped to unite the rest of the world, which appeared more determined than ever to sign the agreement even if the biggest polluter was not along for the ride.

Australia said it would not ratify the Kyoto pact on global warming unless the United States and developing countries got fully involved. Many countries, including Australia, felt strongly that the protocol would not be meaningful without the participation of the United States.

Yet, the Kyoto treaty is the law of the land—and its influence is growing. In the United States, a number of Northeastern states are currently involved in the Regional Greenhouse Gas Initiative. The program provides for a state-level emissions capping and trading program. Greg Nickels, mayor of Seattle, Washington, started a nation-wide effort to get U.S. cities to support Kyoto limitations. At the time of this writing, some 418 cities in 50 states had signed on. On August 31, 2006, California Governor Arnold Schwarzenegger joined with the state's legislature in a bill that would reduce the state's greenhouse emissions by 25 percent by the year 2020. It put California in line with Kyoto, but at a later date. California's greenhouse gas emissions are the twelfth largest in the world. On June 7, 2007, leaders of the G8 summit (including the United States) agreed they would aim to at least halve global CO_2 emissions by 2050.

The momentum of the world favors Kyoto's principles.

The Least You Need to Know

- The Kyoto agreement is the closest the world has come to date on a plan to fight global warming.

- The agreement divvied up the burden of reducing greenhouse gases among the wealthiest industrialized nations.

- Third-world nations were let off the hook for the time being—a point of contention in the U.S. Congress.

- The agreement allowed a system of barter whereby polluting nations could buy credits from nations whose reductions exceeded their requirements.

- Nations could gain credits by reducing greenhouse gases in other countries or by planting forests, which take up carbon dioxide.

- Despite the pullout of Australia and the United States, international momentum is on the side of Kyoto ideals.

Chapter 18

What If It's a Bunch of Bull?

In This Chapter

- A critical debate
- The international controversy
- U.S. politics
- The merits of pollution reduction, energy conservation, and rain forest preservation

The debate over the Kyoto agreement encompasses the basic premises and core arguments of what to do about global warming. Despite the evidence, many still doubt whether global warming is a reality. The evidence, although it looks good on all those charts, still amounts to 1.35°F (.74°C) of warmth. The price tag for the cure is nothing to sneer at. Who's going to pick up the bill? How do we regulate trades? How much of this is politics?

Pros and Cons

Few modern issues are as confusing to the average citizen or elicit as much controversy as the arguments over global warming. People who have no background in science or weather are not afraid to weigh in on a discussion of its impact and legitimacy. Many of the discussions are heated. Scientific meetings on the topic have been known to take on the rancor of a barroom full of sports fans arguing over their favorite teams. "I didn't find that in my model" is a common refrain at international gatherings. So, let's take a look at some of the arguments.

Is Kyoto the Answer?

Proponents argue that if we are going to deal with greenhouse gases, it has to be on a global scale. Individual efforts haven't worked. It took an international effort to address ozone depletion (described in Chapter 19); it will take a similar global effort to tackle global warming. Although there may be flaws in the Kyoto treaty, right now it's the only game in town. Although its scope is minimal, at least it's a start.

Detractors believe that we may need an international approach, but it should include developing areas like China, India, Mexico, and Latin America—where emissions are rapidly rising. These countries are not now subject to Kyoto emission-reduction targets. Also, the United States and Australia need to participate; otherwise, the rest of the world is shouldering an unfair burden.

Is It Worth the Effort?

One side admits some scientific uncertainty about the extent of global warming and its consequences. Still, it seems prudent to take some preventative measures now because the buildup of carbon dioxide in the atmosphere is cumulative. If we wait until its effects become more evident, we may have to live with those effects for two centuries or more.

The other side professes that although carbon emissions may be significantly reduced at relatively reasonable costs, entirely stopping climate change will require 50 to 60 percent cuts. This is likely to be hugely expensive, requiring major reductions in the use of gasoline, fuel oil, natural gas, and coal.

What About Voluntary Measures?

Proponents of voluntary measures argue that business must make emissions reductions gradually. The scope of the changes is not something that can be done overnight. Technologies need to be developed, capital investments planned far in advance, and markets nurtured. Businesses need the flexibility to make those changes over time.

Detractors state that the United States made a commitment to return the nation's emissions to 1990 levels, but in the last fifteen years there has been 20.4 percent emissions growth. Canada made a commitment in 1990 to stabilize its greenhouse gas emissions at 1990 levels by the year 2000. But come 2000, they were actually 20 percent above those levels.

Is the Technology in Place?

Proponents of Kyoto argue that we can go a long way toward meeting emission-reduction targets with existing technologies. But to meet the longer-term goals, we are going to need new technologies. To get there, we need to start making the investments now.

Detractors state that the only way we can meet the Kyoto goals is by accelerating technological development so that we can avail ourselves of new technologies and new fuels. These technologies could be in place within a decade, but that won't be fast enough. To expect that amount of progress within the time frame of the Kyoto agreement is unrealistic.

> **Global Warnings**
>
> The U.S. Department of Energy announced in 2005 that since 1990, the United States has increased its annual emissions of carbon dioxide by an average of 1.2 percent a year. Whereas in 1990, the United States emitted 4.990 billion metric tons of CO_2, in 2005 it emitted 6.008 billion metric tons—a 20.4 percent increase over that 15 years.

Is It Too Great a Burden?

There have been a number of cost-benefit analyses that have tried to answer the question of whether the human and environmental benefits from eliminating or slowing global warming exceed the costs of cutting emissions. The World Resources Institute reviewed 16 of these simulation studies and concluded that reducing carbon emissions to 1990 levels by 2010 would affect the U.S. Gross Domestic Product (GDP) by a range of a 2.4 percent loss to a 2.4 percent gain. Making the recommended reductions of 50 to 60 percent could cost as much as 7 percent of the GDP, which is $530 billion (or $2,000 for every man, woman, and child in the country). Seven percent of the GDP is twice the annual military budget, three times the federal cost of Medicare, and one and a half times the cost of Social Security. That amount of money could go a long way toward ending social problems such as poverty and unemployment.

On the other side, a coalition of environmental groups, including the Union of Concerned Scientists and the Natural Resources Defense Council, published their own analysis, a 20-page document titled "Energy Innovations," which argues that North America has an enormous potential for improving energy efficiency and developing renewable sources such as solar and wind power. They argue that reducing carbon dioxide emissions by 30 to 40 percent can be achieved while saving money, primarily because implementing energy conservation measures is cheaper than using energy wastefully. The National Wildlife Federation states that many of these studies fail to mention the potentially enormous costs of inaction. Economic estimates reach as high as $400 billion a year by the middle of the next century if greenhouse gas emissions continue at the current pace. Those costs include losses to coastal property due to rising sea levels, agriculture due to droughts and floods, and infrastructure due to increased hurricanes and other major storms.

Is It Healthy?

The IPPC predicts that global warming could bring increased floods, droughts, and forest fires to varying places. We only have to look at the tragedy caused by Hurricane Katrina to understand some of the tragic potential that may be in store for us if the weather turns extreme. Disease will spread as ticks and mosquitoes move poleward from the tropics, carrying deadly infections.

Detractors say that's all hypothetical. What Kyoto will most definitely do is increase fuel bills. Higher fuel bills will price air conditioning out of the reach of poorer consumers. While this might not harm young people, older people are more susceptible. In the 1995 Chicago heat wave, some 339 seniors perished from heat exhaustion. Higher fuel costs might turn some to wood-burning fires, which can cause respiratory problems for children.

Then there's the question of mandatory increases in fuel efficiency for cars—a lot of which will come from building smaller, lighter vehicles. Despite air bags and tougher materials, collisions in smaller cars usually mean trouble. In 1991, the National Highway Traffic Safety Administration (NHTSA) concluded that the federally mandated drop in vehicle weights—from 3,700 to 2,700 pounds in the 1970s—boosted highway fatalities by 2,000 annually and serious injuries by 20,000.

But then, a series of high-profile reports emerged about people killed in sport-utility vehicles (SUVs) from rollovers. Sixty-one percent of all SUV deaths involved rollovers. Car crash injuries fell in 2002 while SUV rollover deaths jumped in 2002 to more than 2,400. The NHTSA tested the Chevy Blazer and GMC Jimmy and gave both vehicles a one-star safety rating. The Honda Accord received five stars. According to the NHTSA, SUVs were at least twice as likely to roll over as a passenger car. Increased use of light trucks as substitutes for cars, instead of protecting owners, had produced at least 2,000 additional rollover fatalities annually.

The Price Tag

The IPCC has assessed the global price of stabilizing atmospheric carbon dioxide at twice preindustrial levels by 2100 at around $1 trillion to $8 trillion. Although that sounds like a lot, Steven Schneider—professor of biological sciences at Stanford University—claims it's really minimal. Economists estimate the world will be 10 times richer by 2100. When you add on the costs of tackling global warming, it postpones that target by two years. That means that the world will have to put off being 10 times richer until 2102 instead of 2100. Similarly, meeting the standards set by the Kyoto agreement means that industrialized nations get 20 percent richer by June 2010 rather than by January 2010. According to Schneider, nobody will know the difference.

Hot Debates

Businesses say that to increase the cost of doing business in North America without tax incentives is simply going to move production to developing nations that don't have to adhere to Kyoto or greenhouse gas emissions reductions. Because scientists tell us these gases mix globally, we might lose the business—but the global warming produced elsewhere will come back to haunt us.

Is the Science Sound?

Anti-Kyoto petitions then flooded boardrooms, union halls, and the nation's capitals. One petition known as the "Petition Project" was signed by 15,000 scientists of various specializations (but with no official sponsor), urging the U.S. government to reject the Kyoto treaty based on a lack of convincing evidence. At Congressional hearings, the testimony of a few climate skeptics—many funded by industry—created the impression of considerable climate uncertainty.

On the other hand, the IPCC—not a group of unrelated scientists or industry spokespersons but an expert committee of the world's leading university and government climatologists—has spent more than a decade conducting comprehensive, peer-reviewed assessments of climate change. Its conclusion is that humans are, indeed, influencing the global climate and that climate change is underway. The 2007 IPCC report is more certain of this than the report in 2001.

The International Debate

The international debate takes on different shades of opinion. Countries in the southern hemisphere feel that the problem was created by countries in the northern hemisphere but will affect them more greatly because disease, inclement weather, and environmental effects are expected to hit the southern hemisphere the hardest. Without financial resources, they won't be able to adapt as well, either. The international picture is certainly far from homogenized.

Developing Countries

Members of the Organization of the Petroleum Exporting Countries (OPEC), particularly Saudi Arabia and Kuwait, have been strong opponents of Kyoto. After all, they make their living off fossil fuels. Much of the developing world, which already accounts for around 40 percent of the greenhouse gas emissions, has also been against the agreement.

They feel the rich nations are largely responsible for the problem and that the current level of greenhouse gases in the atmosphere is the product of 150 years of development by those nations. Now that they have theirs, the rich nations want to put the brakes on the growth in the developing world.

Developing nations argue that they need cheap energy to fuel growth. China, with one third of the world's proven reserves of coal, relies on coal for about 80 percent of its energy needs. It plans on expanding coal production fivefold by 2020. Brazil and Malaysia, home to a large percentage of the world's tropical rainforests, are afraid that the treaty might limit their ability to log and export timber or to clear land for agriculture.

Europe

According to many Europeans, part of the problem is that most nations speak of economic growth as the only alternative. But many in Europe feel that the human enterprise might just have grown as big as it can possibly get. The governments of Europe are vastly influenced by their green parties and are determined to push environmental issues. European populations are now stable or shrinking. Constant growth is not the imperative it once was in the Old World.

The Skeptical Environmentalist

Not all Europeans are environmentalists. Bjorn Lomborg, director of the Environmental Assessment Institute in Denmark and author of *The Skeptical Environmentalist*, feels that environmentalists exaggerate the negative state of the environment. They claim that natural resources are running out; that the population is growing, leaving less to eat; forests are disappearing; and the planet's air and water are getting even more polluted.

Lomborg counters that natural resources are increasing, that there is more food now than at any time in the world's history, and that species are becoming extinct—but not as fast as environmentalists claim. He believes that most forms of pollution are associated with the early phases of industrialization.

He thinks that the problems of the world will best be solved not by restricting economic growth but by accelerating it. He believes it's far more expensive to cut carbon dioxide emissions radically than to pay the costs of adapting to increased temperatures. He believes that for what it costs the United States alone to comply with Kyoto, the United States could furnish every person in the world with basic health, education, family planning, water, and sanitation services.

Lomborg claims focusing on development and industry in the Third World rather than conservation will yield the best return in the long run. According to Lomborg, only when people are rich enough to feed themselves regularly do they worry about the effect that their actions might have on their fellow man and future generations.

Lomborg has a supportive following on many talk radio shows but has few supporters in international scientific communities.

Opinion Changes

In May 2002, 100 Nobel Prize winners—including recipients from every year between 1972 and 2001—issued a statement on the 100th anniversary of the Nobel Prize in Oslo, Norway. They claimed that global warming originating from the wealthy few would most gravely affect the fragile ecologies of the poor and the disenfranchised—the majority of which live a marginal existence in equatorial climates. They believe that this injustice represents the most profound danger to world peace in the coming years. They stated, "It is time to turn our backs on the unilateral search for security in which we seek shelter behind walls. Instead, we must persist in the quest for united action to counter both global warming and a weaponized world."

And it's not just scientists who are taking notice. Even as nature crosses its tipping points, the public seems to have reached its own. For years, popular skepticism about climatological science stood in the way of addressing the problem—but the naysayers have become an increasingly marginalized breed. In a new *TIME*/ABC News/Stanford University poll, 85 percent of respondents agreed that global warming probably was happening. Moreover, most respondents say they want some action taken. Of those polled, 87 percent believe the government should either encourage or require the lowering of power plant emissions, and 85 percent think something should be done to get cars to use less gasoline.

U.S. Politics

When 25 nations gathered together in Montreal in 1987 to solve the ozone problem, global atmospheric science was enjoying a respite from politics in the United States. It's true that through the 1960s and 1970s, questions about pollution had pitted the rising environmental movement against some sectors of industry—but it seems that by the 1980s, many of the old adversaries were on the same side. This was evident in the multilateral approach that was taken against ozone depletion.

Conservative Ronald Reagan hailed the agreement as "a model of cooperation." But somewhere along the way, environmental concerns have become an issue of liberal versus conservative. And this is the tragedy. How does an effort to better understand the global climate and the effects that we might be having on it become political? How does it evolve from a "we" issue to an "us versus them" problem?

George H. W. Bush was a sportsman hunter who supported environmental issues, including the Clean Air Act. Bush Sr.'s credit-trading system was used as a model for the credit-trading system in the Kyoto agreement. Bush Sr. was able to cut acid rain in half by allowing U.S. utilities to trade sulfur dioxide credits. At the 1992 Earth Summit, he made the commitment to cut emissions to 1990 levels by the year 2000.

Bill Clinton rode the fence on global warming. He supported the environment but also supported world trade and economic growth. He favored Kyoto, but under his administration—a time of economic prosperity—little was done to reduce greenhouse gas emissions. In fact, greenhouse gases increased significantly during the 1990s. Vice President Al Gore took a dominant role at Kyoto, but Congress never ratified the agreement.

George W. Bush saw too many flaws in the Kyoto agreement. Chief among them was the fact that developing countries like China and India didn't have to make any commitment. Some felt that by pulling out of Kyoto, Bush was simply stating the reality of U.S. politics toward the whole affair. Although Clinton had backed it, nothing had happened. Why deny reality? The force of economics was against it.

Bush announced his own emissions-reduction plan that tied greenhouse gas emissions to the intensity of the economy. To many environmentalists, that was actually an increase disguised as a reduction. Bush increased taxpayer subsidies for coal and proposed legislative rollbacks to the Clean Air Act, which his father had championed.

Al Gore lost the 2000 presidential election, but he went on to lecture on the subject of global warming and eventually wrote *An Inconvenient Truth*. Then he made a movie of his lecture. It won the Academy Award in 2007 for best documentary. The film essentially built on the slide show he'd presented hundreds of times, trying to convince people that global warming was real and we needed to do something about it … now. Coming on the heels of the 2005 hurricane season—a season in which many felt that climate change had played a major role—the public took notice. In a *New York Times*/CBS News poll, an overwhelming majority of those surveyed (90 percent of Democrats, 80 percent of Independents, and 60 percent of Republicans) said that they favored "immediate action" to confront the climate crisis.

In the end, it may be a crapshoot. The IPCC projected in its 2007 report that by 2100 it will get from 2.0 to 11.5°F (1.1 to 6.4°C) hotter than present. If the climate warms only 2.0°F (1.1°C) more, it might not be so bad. We could live with it. It might even be good for your roses. It might even help your vegetables to grow. But if the climate increases 11.5°F (6.4°C), a temperature change nearly as dramatic as that which ended the last ice age would spell major catastrophe: more floods, droughts, and hurricanes. In other words, more Katrinas.

The problem is that many want to forestall action, especially expensive action, based on early indications. But early indications are all we have. The earth has warmed 1.35°F (.74°C). The trouble is, if we wait for further climate change before we act, the changes we inherit will be ours for hundreds of years.

Many liken it to insurance. We take out insurance on our cars and homes in the event of major catastrophes. We don't expect them. We don't anticipate a major fire in our home or a major accident in our cars, but we want to be prepared. Isn't it about time we took out some insurance against global warming?

The Cause

When I first heard about global warming, I thought, "Isn't this neat." I had written a number of articles in major magazines on the problems of preserving open space, saving our wildlife, reducing pollution, energy conservation, and tropical rainforest preservation. I felt then (and feel today) that these are all worthy goals in and of themselves. But global warming brought them all together and gave them global significance. An international agreement against global warming would put the whole world on the side of the environmentalists.

My first reaction was one of skepticism. "This is too good to be true." I knew that many scientists, particularly those working in the environment, would be more likely to buy into it because it would serve their cause. Only after I started studying the evidence did I lose my skepticism. But even if we just forget about all the evidence concerning global warming, are pollution reduction, energy conservation, and rainforest production worthy goals to pursue?

Pollution Reduction

The burning of fossil fuels is the major contributor to increases in greenhouse gas emissions in the atmosphere. But it's also the major contributor to pollution. When we burn gas, oil, coal, and wood, we release a bevy of pollutants into the atmosphere—many of which have greenhouse gas effects. Although carbon dioxide is not poisonous to our atmosphere, many of the other gases associated with it are. If we decrease the amount of fossil fuels we burn, we will decrease those pollutants and increase the purity of our air and water. The European Union estimates that between 43,000 and 180,000 cases of chronic bronchitis and between 15,000 and 34,000 cases of premature death could be avoided in the European Union by 2020 if all the nations in the European Union adhere to E.U. environmental regulations and Kyoto standards.

Energy Conservation

We live in a world run on fossil fuels. We pay huge environmental costs for their extraction. The landscape is pockmarked with oil derricks, refineries, and vast, open pit mines. Tankers like the *Exxon Valdez* off Alaska and the *Prestige* off Spain sometimes break up and leave environmental nightmares. And increasingly, oil is a factor in international warfare. We are fighting a battle over a diminishing resource. If we now begin an aggressive campaign to develop alternative natural resources, we might even have a little oil left over. The technologies are there; all we need is the proper incentives to get going. The best incentive would be a public that is tired of the consequences of our reliance on fossil fuels and anxious to move on to the next phase.

Rainforest Preservation

Although about three quarters of the increase in greenhouse gas emissions are due to fossil fuels, one quarter is due to deforestation and the clearing and tilling of the landscape. Europe has already cut down most of its natural forests. North America has destroyed much of its own. The biggest virgin forests in the world range over the tropics in South America, western Africa, and southeast Asia. According to Wildlife Conservation International, these are hotspots of yet-to-be-discovered wildlife species. But these forests are in danger. About 80 to 90 acres of tropical rainforest are destroyed every minute. Preserving these forests is a worthy cause. But unless the wealthy nations of the world are willing to help the Third World, the forests will disappear—and so will the enormous wealth of wildlife they contain.

The Least You Need to Know

- Voluntary measures to control greenhouse gas emissions have thus far proved ineffective.

- There are legitimate pros and cons to the efforts to do something about greenhouse gas emissions.

- U.S. politics has gone back and forth over the issue despite global concerns.

- Pollution reduction, energy conservation, and rainforest preservation are worthy goals in themselves.

- Despite the skepticism, it might be time to take out some insurance against global warming.

Chapter 19

Is That a Hole Up There?

In This Chapter

- ◆ The hole above the Antarctic
- ◆ How the ozone layer protects us
- ◆ How we almost destroyed the ozone layer
- ◆ The world bans CFCs

Global warming is not our first climate crisis. In the 1980s, the world fretted over a different problem in our atmosphere: the Antarctic ozone hole. But instead of dividing the world, it drew nations together.

A Startling Discovery

In 1985, the British team led by Joseph Farman discovered a large hole in the ozone layer over Halley Bay Station in Antarctica. The ozone layer is a thin region in the upper atmosphere where ozone (O_3) resides and blocks out deadly doses of solar ultraviolet (UV) radiation. When the team first measured ozone above the station, they assumed that the spectrophotometer they were using must be broken and sent the device back to England to be repaired. But the spectrophotometer worked just fine.

In the summer of 1986, NASA sent National Oceanic and Atmospheric Administration (NOAA) senior scientist Susan Solomon and a team of 16 scientists to Antarctica to investigate. Solomon was only 30 years old at the time but had already co-authored a standard textbook in her field and had received an award from the American Geophysical Union.

For more than a decade, scientists had speculated that commonly used industrial compounds known as chlorofluorocarbons (CFCs) were destroying the ozone layer and making humans and nature susceptible to incoming and dangerous UV radiation. Solomon's breakthrough was the discovery of a critical link between manmade CFCs and the ozone hole above Antarctica. Solomon's theory was that the ice crystals in the clouds over Antarctica might assist the breakdown of CFCs into reactive chlorine, which destroyed the ozone. It came to be known as the Antarctic Ozone Hole throughout the world from a series of NASA satellite photos. The public began to realize humanity's enormous capacity to affect its environment. With the Antarctic Ozone Hole, we had created a huge, new, semi-permanent feature on the planet that was bigger than the United States. In 1987, the industrialized nations gathered in Montreal to discuss the crisis. In short order, they signed an agreement to set limits on the production of CFCs and other related chemicals.

How did this happen so quickly, smoothly, and with nonpartisan participation? The ozone hole and the way it was handled is a lesson in cooperative international agreements.

The Layers of the Atmosphere

In Chapter 3, we explained how plants in the sea first started creating oxygen. Well, a portion of that oxygen rose to some of the highest regions of the planet's atmosphere, where UV radiation from the sun caused a reaction that led to the forming of ozone. That ozone settled into a thin layer that effectively blocked out deadly doses of solar UV radiation, which until then had bombarded the surface of the earth. It was only with this atmospheric shield that animal life, which first developed under the protection of water, was able to crawl out on land. To understand the science and the controversy surrounding the ozone layer, you have to understand a little bit about the multiple layers of our atmosphere and how sunlight reacts as it passes through them.

The *thermosphere* is the top layer of the atmosphere. It extends up to 400 miles or more above the surface of the earth. As the sun's rays enter this level, the UV portion of the spectrum interacts with the atmosphere, which at that height is pretty thin. But what's up there quickly grabs the heat in those rays, which is why the thermosphere is quite hot at higher altitudes. It cools, however, as it descends toward the *mesosphere*.

Although the temperature in the outer limits of the thermosphere can heat from UV radiation, the air there is so thin that the fast-moving air molecules—although hot— seldom collide with anything. If an astronaut were to stick his or her hand out into the hot part of the thermosphere, his or her hand would not feel hot.

As the sun's rays hit the mesosphere (the next layer of our atmospheric "cake"), the temperature reverses and starts rising again. Although the air is still thin here, there is enough to absorb the rays and restart the heating process.

Below the mesosphere is the *stratosphere*. In the stratosphere, we start getting cold again. But the drop in temperature isn't as big as it is through the thermosphere. The weather through the stratosphere is closer to that of the earth's surface.

As the sun's UV rays finish their descent through the stratosphere, they encounter the *troposphere*. That's where we live. It's the bottom of the atmosphere, and it extends from sea level to about 7.5 miles (12 kilometers) high. When the sun's UV rays hit the troposphere, the air is pretty cold—less than -60°F (-51°C). But the UV rays of the sun hit the surface of the earth and radiate back infrared radiation (heat). This is where the greenhouse gases do their good work by absorbing that heat and making things comfortable. The troposphere is also the layer where all our weather takes place. It's generally where the clouds stop. Seldom can they penetrate the stable air of the stratosphere (except in violent storms).

def•i•ni•tion

The **troposphere** is the bottom layer of the atmosphere, rising from sea level up to an average of about 7.5 miles (12 kilometers). The **stratosphere** is above that and rises to an average of 31 miles (50 kilometers). The **mesosphere** lies on top of that and runs upward to about 50 miles (80 kilometers). The **thermosphere** tops the atmosphere off, rising to around 400 miles (644 kilometers)—although it really has no well-defined upper limit.

The Ozone's Vital Role

The greatest absorption of the UV rays takes place between 25 and 50 kilometers above the earth. The peak concentration of ozone occurs at around 25 kilometers, the infamous *ozone* layer. The ozone layer is in the stratosphere—the layer above the troposphere where we all live. The stratosphere contains about 97 percent of all the ozone in the atmosphere. Oxygen (O_2) rises from the earth's surface up into the stratosphere, where it reacts with the sun's incoming UV rays, goes through a series of chemical changes, and comes out as ozone (O_3). This layer of ozone encounters the

UV rays, breaks up, and reforms in a continual process that absorbs the UV rays. It's like a celestial pool game with one ball connecting to another and then another—each connection taking up more UV rays.

This chemical pool game protects us from the extreme UV rays that enter the stratosphere. We still get some UV rays but not the dose that we would get without the ozone layer. In fact, if those UV rays passed through the stratosphere unimpeded, they'd take us out.

One of the reasons why scientists feel that Mars is lifeless today is because it has lost its UV shield, and that makes the place currently uninhabitable. UV radiation there bombards the planet's surface, creating reactive superoxides and peroxides that kill any available organic molecules. As Carl Sagan put it, "The surface of Mars is antiseptic with a vengeance."

Billions of years ago, there was no ozone layer to protect the earth from the sun's UV radiation. Scientists believe that life only existed in the sea, protected by the water. However, a layer of ozone gradually developed in the stratosphere. It was only about 600 million years ago that this layer grew thick enough to trap most of the UV radiation and the land could then support life.

How good ozone is made.

A Strange-Smelling Chemical

Ozone was first discovered in 1840 by a German chemist, Friedrich Schoïnbein. It's said that he was working in a poorly ventilated laboratory and became conscious of a peculiar odor near electrical equipment. He tracked it down and discovered a gas, which he called *ozone*, which is from the Greek word for "smell." Thomas Andrews, an Irish chemist, later identified it as a form of oxygen.

Its presence in the upper atmosphere was deduced at first theoretically by the absence of certain wavelengths of solar UV radiation at ground level. Scientists deduced the absence could have only occurred if the radiation were intercepted by ozone. Later, rockets and satellites confirmed this initial finding.

The first estimates of the abundance of ozone in the atmosphere were made in 1913. Systematic observations all around the world began in 1930. But it wasn't until 1974 that the world's attention was focused on the ozone layer. Then, chemists F. Sherwood Rowland and Mario Molina of the University of California at Irvine announced to the world that the ozone layer was under attack by a family of commonly used industrial compounds known as chlorofluorocarbons (CFCs).

The Ozone Depletion Story

CFCs were a family of industrial compounds created in 1928 by an eccentric Ohio scientist named Thomas Midgley Jr. CFCs are the chemicals that make your refrigerator work, and in the days when everybody was still waiting for the iceman to come, that discovery was a big deal. CFCs were thus the miracle chemical of the day. Plus, CFCs had a lot of other uses as well. They were the forces behind aerosol sprays. They were used as fillers in insulation. They were also used in air conditioning units to cool homes and cars.

The reason why CFCs did their job so nicely was that they were stable chemicals. Nothing scientists had tried yet could break them up into their constituent atoms of chlorine, fluorine, and carbon. Before the 1970s, these chemicals were thought to be harmless. But the guys who said they were harmless were wrong.

In the early 1970s, Mario Molina—a chemist who'd worked on lasers at the University of California at Berkeley—joined a group under F. Sherwood Rowland at the University of California at Irvine. Molina's new job was to track down CFCs that had been accumulating in the atmosphere for some time but were disappearing recently. His first task was a systematic search for processes in the lower atmosphere that might destroy them—but CFCs were so tough, nothing seemed to affect them. So the question was, where had they gone?

Mario Molina was born in Mexico City. He attended the University of California at Berkeley in the 1960s, where he earned his Ph.D. and worked with chemical lasers. He was dismayed by the fact that they could be used for military purposes, as he'd been a peace activist at Berkeley. So when he moved to Irvine, California, and joined the group of F. Sherwood Rowland, Molina chose as his research project the benign goal of finding out the environmental fate of certain inert industrial compounds called CFCs, which had been accumulating in the atmosphere for some time. Molina had no idea of the controversy he was about to encounter.

Molina and Rowland determined that the chemicals would eventually drift into the upper atmosphere, where they could be destroyed by solar radiation. The problem was that the chlorine atoms produced by the decomposition of CFCs could destroy ozone. The continued release of CFCs into the atmosphere could cause a significant depletion of the ozone layer—the chemicals in the stratosphere that protected Earth from excessive UV rays.

The two published their findings in the June 28, 1974, issue of *Nature*. But they didn't stop there. Molina and Rowland knew that the information could die if its only exposure was to the scientific community through their paper. So in the years following publication, the two scientists took to the road and the air to spread the message to policy makers and the news media. They continued to publish articles on the CFC/ozone issue and presented their results at scientific conferences and testified before legislative hearings where lawmakers were considering controls on CFC emissions.

They projected a decrease in the ozone content in the atmosphere that could have serious health effects. These gases had been around for 50 years, so a lot of damage had already been done—and if we didn't stop putting them into the atmosphere, it was going to get worse. And even if we ceased manufacturing these gases completely, it would take another 100 years for the atmosphere to return to preindustrial conditions.

Molina and Rowland set into motion a contentious debate between scientists, policy makers, environmentalists, and industry leaders that was to last for 13 years. CFCs in aerosol sprays were banned in the United States in 1978, but many people remained unconvinced of their dangers. Why eliminate the use of a vital chemical and a multimillion-dollar industry based on hypothetical evidence?

The Price of Too Much UV

So what is all the fuss about? Molina and Rowland weren't talking about an end to the ozone layer; they were talking about a reduction of 3 to 7 percent. What's that?

A little more UV radiation? Don't people go to tanning salons for that sort of stuff? But even this level of ozone depletion would bring serious health consequences.

Effects on Humans

According to the United Nations Environment Program, increased UV radiation can impact our immune system and lead to the spread of infectious disease, increase the risk of skin cancer (particularly basal and squamous cell carcinomas and the more deadly melanoma), and cause cataracts. Increased UV radiation affects our immune system responses, skin, and vision.

The evidence continues to mount that UV radiation can suppress immune reactions in both laboratory animals and humans of all skin types. UV radiation can impair certain antibodies. Studies show that UV radiation can not only decrease the immune responses to infectious organisms but can also increase the severity and duration of a number of infectious diseases. And not only that—it can also decrease the effectiveness of vaccinations. Furthermore, UV radiation can seriously impact Earth and its aquatic ecosystems.

UV rays are also known to cause skin cancers. However, a recent study estimated that in the absence of the Montreal Protocol, an international agreement curtailing CFC production, elevations in UV radiation due to ozone depletion would cause a fourfold increase in skin cancer by the end of the twenty-first century. Even under current restrictions, the relative increase in skin cancer is still estimated to increase by 10 percent until 2060. After that, it should gradually return to preindustrial levels. That requires everyone to follow the rules, though.

Recent experiments show that rats that are exposed to UV rays develop the precursors to cataracts at the periphery of the lens.

Effects on the Earth

UV radiation can have serious effects on plant flowering, pollen development, seed production, and seed size. UV rays can also promote plant pests, including various microorganisms and insects. It can even damage the plant's DNA. Plants exposed to excessive doses of radiation show a variety of deformities like shorter stems, multiple branching, and other effects.

There are some positive effects, however. Some Mediterranean species developed thicker wax layers on leaf surfaces under elevated levels of UV radiation, which helps the plants retain more water during the dry summer months.

Effects on Aquatic Ecosystems

Oregon State University researchers discovered strong evidence in the 1990s that the decreasing number of amphibians, especially toads and frogs, throughout the world is related to an increase in UV radiation. The researchers hypothesized that the increased UV radiation destroys amphibian eggs. Serious declines in the world's coral reefs may also be an effect of increased UV rays.

Phytoplankton are tiny plants found in the top sunlit portion of the ocean that drift with the currents. They are the basis of most of the ocean's food chain. Krill eat phytoplankton and a number of animals eat krill, including penguins, seals, and humpback whales. Although UV radiation might not harm humpback whales directly, it will kill phytoplankton—and that will destroy an important part of the food chain upon which the whales depend.

def•i•ni•tion

Phytoplankton are microscopic plant life found drifting in the upper sunlit portion of the sea or freshwater and are the basis of the food chain for other marine organisms.

Phytoplankton are also a major consumer of atmospheric carbon dioxide. Thus, the destruction of phytoplankton could have immediate effects on wildlife as well as lead to increases in global warming.

The Antarctic Ozone Hole

In the early 1980s, a lot of this was still speculative. Then, British scientists discovered a large hole in the ozone layer—and Susan Solomon and a team of 16 scientists confirmed it. The world woke up to the fact that CFCs were destroying the ozone layer.

Climatoids

When CFCs are broken down in the atmosphere, they liberate chlorine atoms. One chlorine atom is capable of destroying 100,000 ozone atoms.

Scientists got busy checking elsewhere. A series of careful measurements taken from old converted U-2 spy planes as well as the Nimbus-7 satellite indicated that ozone depletion over North America and Europe was taking place at a much faster rate than was previously anticipated. Declines were as rapid as 0.5 percent a year. That figure meant that the time when ozone depletion could have serious effects on health and the environment was shortened from a century to decades.

Global Warnings

By the mid-1980s, scientists were also uncovering problems with greenhouse gases. They discovered that CFCs were not only dangerous to the ozone layer but were also an efficient greenhouse gas. Just like carbon dioxide, methane, and nitrous oxide, CFCs contributed to global warming by absorbing some of the infrared heat reflected back from the surface of the earth. With the bad press mounting, it seemed that CFCs were doomed.

The World Fights Back

In 1987, the industrial nations gathered in Montreal to discuss the crisis. Although there were still members of the chemical industry that urged restraint, few of the international representatives were willing to take the risk in light of the mounting evidence. In the end, 25 nations signed the agreement to set limits on the production of CFCs and other related substances. By that initial agreement, there was to be a gradual phase-out of CFCs and other ozone-depleting chemicals. It was the first global environmental accord. Today, more than 168 nations are a party to that agreement.

On April 5, 1988, President Ronald Reagan issued a statement praising the agreement: "The Montreal protocol is a model of cooperation. It is a product of the recognition and international consensus that ozone depletion is a global problem, both in terms of its causes and its effects. The protocol is the result of an extraordinary process of scientific study, negotiations among representatives of the business and environmental communities, and international diplomacy. It is a monumental achievement."

The initial agreement called for a gradual phase-down of production and consumption of CFCs and related chemicals, but as the bad news continued to mount from the scientific arena, a sense of urgency grew. At subsequent meetings in London, Copenhagen, Vienna, and again in Montreal (all in the 1990s), the phase-out was accelerated. Most CFCs were banned by 1996.

Today, scientists have developed a number of alternatives to CFCs, and more are being developed. Scientists report that the effort appears to be worth it. The ozone layer is already showing signs of recovering. Special instruments aboard satellites have shown that the amount of chlorine in the stratosphere peaked in 1997 and appears to be on the decline. The ozone hole over Antarctica grew to its maximum size in 1998 and was a little smaller in 1999. It appears to be oscillating around this high value but is expected to go down. Scientists predict that it may not be until the middle of the twenty-first century before the effects disappear.

Rowland and Molina, who were initially hounded for their predictions in the mid-1970s, were vindicated when they received the Nobel Prize for their discoveries. It was the first Nobel Prize ever given for an environmental science.

Scientists Look Elsewhere

Perhaps the biggest prize of all was the global cooperation that this issue seemed to engender. Despite extreme opposition from the chemical industry, lots of controversy in the scientific community, and the challenge of fashioning an agreement that would satisfy the needs of both the industrialized and nonindustrialized worlds, everyone came together and negotiated an agreement that for the first time tackled an important environmental issue.

Under the excitement of this important moment, many now looked to the problem of global warming and hoped for similar cooperation and a similar result. But the going hasn't been as smooth. Banning CFCs involved controlling the production of a single family of industrial chemicals, but the causes of global warming are much more numerous.

In the end, it's gotten down to the fact that only a few dozen companies worldwide produced CFCs. According to one negotiator, you could put them all in one room and talk to them. But you can't do that for the producers of greenhouse gases. Imagine the size of a room that would hold representatives from all the world's utilities and industries. Carbon dioxide, methane, and the other greenhouse gases are the by-products of the very heart of modern civilization—including industry, transportation, power, and agriculture. It's a much bigger ballgame. Still, the way the world attacked the problem is a model of how the world once tackled a major climate problem … and how it might do the same with global warming.

Ozone Down Here

Ozone in its natural form is not only present in the stratosphere but also at ground level, mostly from ozone manufactured in the stratosphere that has floated down. But in the twentieth century, another type of ozone has emerged. In the presence of hydrocarbons and nitrogen oxides coughed from the exhaust of automobiles, ozone is produced through a series of complex chemical reactions that lead to smog.

The origins of the word *smog* may have come originally from a combination of fog and smoke in Old England, but more recently it has come to mean the polluted haze found in many American cities—particularly in the southwestern part of the United

States. It's not really smoke and fog but a chemical stew of pollutants (primarily from automobiles) that are transformed to smog by a reaction with the sun.

Smog's primary ingredients are nitrogen oxides and volatile organic compounds, which create ozone when mixed together. That's right—the ingredient that's so vital to humankind up in the stratosphere is real nasty when encountered down here on Earth. There are a number of other chemicals within smog that are dangerous, but ozone seems to get all the bad press.

def•i•ni•tion

Smog is pollution and comes in two types. The gray smog of older industrial cities like London and New York is derived from the massive combustion of coal and fuel oil in or near the city. The brown smog characteristic of Los Angeles and Denver in the late twentieth century comes from automobiles.

Elderly people in Los Angeles, Denver, and other southwest metropolitan cities affected by smog frequently go to malls early in the morning to walk or exercise. The air-conditioned climate protects them from the damage of ozone and is supported and encouraged by businesses and communities alike.

Ozone can irritate the eyes, nose, and throat. Ground-level ozone poses a significant health risk to people who suffer from chronic lung disorders such as asthma and bronchitis. Lungs will age a lot faster in an environment with ozone, even at low levels. It's also bad if you exercise, work, or engage in recreational activities outdoors.

Ozone is not only bad for your health but it can also ruin your tires, take the shine off your paint job, make your nylons run, and wreck your clothing. Plus, it's nasty to crops and plants. Some people estimate the damage to crops in the United States at about $1 billion a year. It also attacks the cool, moist leaves in trees and can contribute to forest fires.

Ozone in the Artic

To top it off, ground-level ozone is a big problem in the Arctic. According to new research by NASA's Goddard Institute for Space Studies in New York, ozone was responsible for one third to half of the observed warming trend in the Arctic during the winter and spring. Ozone is transported from the industrialized countries in the Northern Hemisphere to the Arctic quite efficiently during those seasons. Not only is ozone a problem in the upper atmosphere, down here at lower levels it also acts as a greenhouse gas and contributes to global warming.

The warming impact of low-altitude ozone on the Arctic is very small in the summer months because ozone from other parts of the world does not have time to reach the region before it is destroyed by chemical reactions fueled by ample sunshine. Reducing ozone pollution can not only improve air quality but also have the added benefit of easing climate warming, especially in the Arctic.

The Least You Need to Know

- Our atmosphere is a multi-layered affair.

- Ozone in the stratosphere protects life on Earth from a nasty dose of ultraviolet radiation.

- Scientists in the 1970s discovered that the ozone layer was under attack by CFCs, which were commonly used for air conditioning and refrigeration.

- The world got together in Montreal and agreed to phase out CFCs and their sister chemicals.

- Ozone may be beneficial in the stratosphere, but down here on Earth it's not only unhealthy but may also contribute to global warming.

Chapter 20

Summing Up the Argument

In This Chapter

- ◆ The historical evidence
- ◆ The role of carbon dioxide
- ◆ The havoc global warming is playing now
- ◆ What global warming will do in the future

Global warming and climate change are complex topics that bring together a number of different sciences. The issues are vast and complex. The fix is costly. The evidence is murky, and there is a lot of speculation. In this chapter, I will try to restate the case succinctly, review the most important facts, state the most current evidence, and discuss the scientists, their science, and their predictions.

Unraveling the Climate

I'm standing in the Laboratory for Tree-Ring Research under the football stadium at the University of Arizona in Tucson. Thomas Swetnam, the bearded director of the center who studies tree rings for clues to past climates, holds a piece of wood from a dead giant sequoia tree that is about three feet long, four inches wide, and a half inch thick.

It's a section from the giant sequoia that was cut from the bark to the center of the tree. It displays the annual rings of the tree as half moon lines spaced tightly together. On this piece of wood are pins with little quarter-inch flags that have dates on them. The flags start at 1958 C.E. and go all the way back to 523 C.E. Swetnam points to a section of the rings in the center of the board that are labeled as the late 1300s. Here, the rings are closer together than elsewhere on the board.

Global Warnings

Scientists at the Laboratory for Tree-Ring Research are able to tell a lot about both temperature and precipitation by the width and density of the tree rings. Trees generally put on fatter rings when it's wetter or warmer than normal. Reading rings of a tree at the highest place that they grow on a mountain will tell you how the temperature varied. Reading the rings from a tree at the lowest place they grow will tell you precipitation levels.

"That's what got the Anasazis," says Swetnam. He explains to me that from about 800 to 1200 C.E, the Anasazi Indians of the Southwest desert built elaborate cities in a number of places in Arizona, New Mexico, and Colorado. "But then, in the late 1300s, they ran into what the tree rings tell us was a long drought. They had to abandon those fabled cites and move south along the Rio Grande River."

The Laboratory for Tree-Ring Research was originally developed back in the 1920s to date these great Indian ruins, and scientists found that they could create a calendar by matching the tree rings in the wooden beams of the ruins. Now the laboratory has joined with other scientists who are looking at a number of past climate signals, including tree rings, ice cores, and marine sediments. They're trying to unravel another human mystery as to whether we are headed for a similar disruption in our society as the Anasazis encountered—only this time due to global warming.

The Common Belief

Most scientists now believe that humans are changing the climate of the planet by their greenhouse gas emissions. The controversy is over the amount we're changing it and over the high costs of repairing the damage. Recently, scientists at the Laboratory of Tree-Ring Research put together a compilation of global temperature estimates of the last 1,000 years. The estimates were derived from a number of sources, including thermometers, tree rings, ice cores, and marine sediments. The resultant graph from all these disciplines looks like a hockey stick, where the end of the graph depicts a sharp rise that represents the increase in temperatures over the last 150 years.

This chart is used by the IPCC in its reports on global warming. More than 2,000 scientists put together the 2007 report based on almost two decades of research. Although not every scientist signed off on every word, they did reach a consensus. They estimate that if the current rise of carbon dioxide in the atmosphere continues, the world will warm up from 2.0 to 11.5°F (1.1 to 6.4°C) hotter than present by the end of the twenty-first century. The resultant hurricanes, floods, droughts, and sea level rise could drown much of Florida, trash our agriculture, and cause extensive damage and disease.

Historic Gases

In 1827, French mathematician Jean Fourier discovered that certain gases in the atmosphere act like a greenhouse. They let in sunlight like a greenhouse does but prevent some of the sun's warmth from radiating back into space.

The greenhouse effect is the consequence that certain gases have on the earth's climate—chief among them carbon dioxide. Although they constitute less than 1 percent of the atmosphere, they act like a blanket covering the earth. Without the greenhouse effect, the average global temperature would be around 0°F instead of the current toasty 59°F.

Not until the 1930s did anybody notice that there were problems with what scientists had come to call the greenhouse effect. Then, during a period of particularly hot summers in Europe, George Callendar—a British coal engineer—discovered that the world was gradually getting warmer. He pointed to rising levels of carbon dioxide as the culprit.

Charles Keeling was the first to develop an instrument for measuring CO_2 in the atmosphere. In 1958, during the International Geophysical Year, he took it up to the top of Mauna Loa, a high volcano on the island of Hawaii. There, he recorded the rise and fall of CO_2 in the atmosphere throughout the seasons as plants grew and decayed. He left his instrument up there, put a few more out in other parts of the world, and discovered that CO_2 was gradually increasing in the atmosphere.

Prior to the Industrial Revolution (which began in England in about 1750), the concentration of carbon dioxide in the atmosphere stood at about 280 ppm. In 1958, when Keeling first measured it, it was 315 ppm. In 2005, it was 379 ppm. Scientists expect that the concentration of carbon dioxide in the atmosphere will double over preindustrial levels to 560 ppm sometime during the twenty-first century.

Murky Thermometers

Scientists then started looking at the temperature to see whether there was a cor-
relation. It wasn't easy. Historical temperature records stretch back 300 years in a
few places but only a few decades in most. Even in places where we've been keeping
records, if you put your thermometer in the middle of a town and a city grows up
around it, the asphalt and concrete buildings make it
warmer. It's called the "urban heat island effect."

Satellites have made the readings of the earth's temperatures more accurate, but historical records are
lacking. To get an idea of what temperatures were
like in the past, scientists have had to look at certain
proxies—such as tree rings, ice cores, and marine sediments—that they can analyze and extract readings of
historical temperatures.

def•i•ni•tion

Proxies are things like tree rings,
ice cores, and marine sediments
that scientists can analyze for
indirect readings of historical temperatures of the atmosphere.

Calibrating the Computers

One of the main reasons for studying the past is to help scientists predict the future
of the atmosphere. Scientists use basic computer models for determining the weather
and turn them toward the distant future. To do this, the programs are simplified so
that instead of telling us weather patterns in specific places, they tell us general climate parameters (such as temperature). This is necessary because weather programs
test the limits of today's supercomputers just to predict the weather 10 days ahead. To
give us a picture of the climate 100 years out, you can only ask the computer simple
questions.

Scientists have had some luck predicting El Niños almost a year in advance, but
predicting climate 100 years out is untested science. To verify their accuracy of the
models, scientists run their computer programs on past climate parameters and see
whether they predict the known outcomes. Our studies of past climates help us calibrate the computers that we then point toward the future.

Tree Rings and the Climate

At Tucson, they've built a chronology of tree rings from pencil-width cores taken
from the bristlecone pines in the White Mountains and the giant sequoias in the
Sierra Nevada Mountains. These trees are some of the oldest on earth. One calendar
that the lab put together of bristlecone pines went back 9,000 years.

During those years, a number of climate events have affected man. A warming trend from about 950 to 1250 C.E. may have been a welcome respite in the United States, but it also made it possible for the Vikings in the Old World to rape and pillage Europe. The Vikings discovered America at that time and established a settlement in Greenland. But when a cooling trend known as the Little Ice Age came along about 1400 C.E., Viking settlers in Greenland brought their farm animals into their homes. When winter got too cold, they ate the farm animals. They eventually died out when no boats could land on their frozen shores.

The changes in the climate over the last 2,000 years—although they were rough on the Vikings—have represented slow, one-degree shifts in global temperatures. For the last 10,000 years, the climate has been fairly stable. Prior to that, it was a different picture.

Ice Cores and Ocean Currents

To look back farther into the history of the atmosphere, scientists have gone to both Greenland and Antarctica, where they've uncovered a record of both the temperature and the atmosphere buried in the ice. Although a portion of the world's permanent ice is locked into glaciers in high mountain regions around the world, fully 99 percent of the earth's ice is in the great ice sheets that cover most of the island of Greenland and nearly the entire continent of Antarctica.

From 1989 to 1993, U.S. and European scientists spent five summers drilling a 2-mile (3.2 kilometers) core into the ice at the center of Greenland to get readings of the atmosphere over the last 110,000 years.

Climatoids

Snow in Greenland doesn't melt but accumulates on the surface and compacts under the weight of new snow accumulating above. Each annual layer is about 12 inches deep toward the top of the ice sheet and about 4 inches deep at the bottom of the 2-mile (3.2 kilometers) ice core. The low-density, coarse-grained layers formed under the summer sun show up as light bands, while the fine-grained snow packed by winter storms appears darker. Scientists can read these cores on a light table.

Inside the ice cores are samples of the ancient atmosphere. The air in the ice accumulates into bubbles and is further compacted but upon release is a fairly accurate representation of the atmosphere at the time the ice was formed.

Scientists determine the temperature at the time the snow fell by measuring the amount of heavy oxygen (oxygen with two extra neutrons) in the compacted snow. It's the first thing to fall out of the clouds when it gets cold, and the amount of it in the ice is directly proportional to the temperature. The bubbles can tell you what was in the atmosphere when it was formed.

The winds of the world do a good job of mixing the atmosphere. If you were to release a large quantity of gas in Seattle, it would only take a few years before people in Bangkok could measure what you've released. Dust and sea salt don't mix well, but most gases are in the atmosphere long enough to mix globally.

The Greenland ice cores give us a marvelous record of the atmosphere that stretches back 110,000 years. What they show is that for the last 10,000 years, the global climate has been very stable. The globally averaged temperature has remained about 58°F (14.4°C), except for the last 150 years when it rose to 59°F (15°C). The Little Ice Age that froze the Vikings out of Greenland was only about 1°F (.5°C) cooler. The present amount of global warming has made us about 1.35°F (.74°C) warmer.

We are currently in an interglacial—a warm period between ice ages. The cycle has occurred every 100,000 years for some time now. Over the 110,000-year record of the Greenland ice cores, there have only been two stable periods: the one we're in now and the one that occurred toward the middle of the last ice age from 75,000 to 58,000 years ago, when central Greenland was about 40°F colder than today.

The rest of the time, the earth's climate was a roller coaster—swinging wildly back and forth. You step out your door one year and it's southern California. A decade later it's Vancouver. Another 10 years and it's Acapulco. From the end of the last Ice Age until the present, it's been really stable except for a couple of dips.

The Younger Dryas

The best-studied dip was the Younger Dryas event. It occurred about 12,800 years ago. The Younger Dryas event is named after the dryas plant, which is a little rose found only in the tundra. During the Younger Dryas, the tundra reached all the way south into Virginia. Back then, the world was warming up out of the last ice age rather nicely when all of a sudden it took a big dive. Europe was about 12 to 16°F (7 to 9°C) colder than today—and it stayed that way for 1,200 years. Scientists wonder whether global warming could trigger another such event.

The Younger Dryas was hell on nature. During the event, the mastodons, mammoths, horses, and saber-toothed tigers all went extinct. Early man contributed to this extinction through aggressive hunting, but scientists feel the climate was a major factor as well. When the Younger Dryas ended, it did so in a series of jumps—with central Greenland warming up 15°F in a single decade or less.

Scientists believe that the Younger Dryas event may have been caused by a shut-off of the Thermohaline Circulation (THC), a conveyor belt–like current in the deep sea that connects the major oceans of the planet. In this current, warm, salty Atlantic water near the equator moves north toward Greenland and Labrador, where it has cooled enough to sink. The warmth that this current brings is the reason why Britain is as warm as it is, although it is in the same latitudes as Canada.

The sinking occurs in a couple of relatively small places in the sea, diving more than a mile deep before the water flips around and heads south down the Atlantic. The high salt content is necessary if the waters in the current are going to make the dive into the North Atlantic and turn south. Scientists feel that global warming could create enough ice melt and rain to dilute the water in the current and shut off the circulation. If it did, Europe and other parts of the northern hemisphere could get quite cold while the southern hemisphere could get hotter.

Scientists believe that the climate might be like a flip switch as opposed to a dimmer. If you put pressure on the switch gradually, it doesn't immediately start to change; it hesitates until there is enough pressure and then flips all of a sudden into a completely different state. Once there, it might just remain for centuries. It's happened before.

The Rise and Fall of Carbon Dioxide

Carbon dioxide might be a critical part of that switch. The Greenland ice cores give us a picture that goes back 110,000 years, but cores drilled in Antarctica give us a picture that goes back more than 400,000 years. What Antarctica has shown us is that during the last four glacial cycles, the atmospheric concentration of carbon dioxide has varied from 200 ppm during the ice ages to about 280 ppm during the present interglacial period. Over the last 10,000 years, it has remained steady at 280 ppm until the Industrial Age, when it started going up. It's currently about 370 ppm.

What does that say? Well, the difference between the ice ages and the interglacials is only 80 ppm of carbon dioxide. Although scientists don't believe that carbon dioxide caused the ice ages, it amplified the effect.

Snowball Earth

About 600 million years ago, an ice age got out of hand and froze the entire planet during a period called Snowball Earth. It seems that once ice starts forming, it may perpetuate itself by reflecting sunlight. The entire ocean may have frozen. Life, which was pretty microscopic back then, was nearly wiped out.

Hothouse Earth

But carbon dioxide kept accumulating in the atmosphere—put there by spewing volcanoes. It increased 1,000-fold over 10 million years, and Snowball Earth turned into Hothouse Earth, where the average global temperature soared to more than 122°F (50°C).

What all this shows is how important the content of carbon dioxide in our atmosphere is to climate.

The Current Picture

So … what do we have now? What we've got is a period during which humans have affected the atmosphere on a more rapid scale than has ever been known in nature. According to the World Meteorological Organization, 9 of the 10 warmest years on record have occurred since 1990.

Although scientists predict that the biggest stuff is yet to come, some changes are already evident. There has been a widespread retreat of mountain glaciers outside the poles just in the last century. Ice in the European Alps has declined by half. There has also been a 40 percent decline in late summer Arctic sea ice thickness.

When naturalists first hiked through Glacier National Park more than a century ago, there were about 150 glaciers nestled into its high cliffs and jagged peaks. Today, there are only 35. In another 30 years, scientists estimate even those will be gone. Glacier National Park will be glacierless.

The United Nations Environmental Program recently completed a survey of glaciers in the Himalayan Mountains and found that dozens of lakes were so swollen from melting glaciers that they could burst in the next few years, inundating villages throughout the region. The collapse of the Maili glacier on the northern edge of the Caucasus Mountains ripped out trees and tossed massive trucks into the air like toys. It buried a Russian village under 3 million tons of ice and mud and left at least 9 people dead.

Four Antarctic ice shelves have collapsed—including the Wordie Ice Shelf, which measured 772 square miles in the 1940s but has shrunk by two thirds today. The Prince Gustav Ice Shelf disappeared altogether. The most dramatic of these collapses has been a series of breakups that began in 1989 on the Larsen Ice Shelf.

As scientists watched via satellite in January 1995, a large portion of the Larsen disintegrated over the span of five days. At one point, a slab of ice the size of Rhode Island shattered free and floated away. On January 31, 2002, an even larger piece called the Larsen-B ice shelf began to disintegrate—and within 35 days (most of it in just two days), an area that measured 150 miles long by 30 miles wide completely broke up. Scientists tell us today that hundreds of glaciers on the Antarctic Peninsula are flowing faster, further adding to sea level rise, and that global warming is the most likely cause.

Over the last 50 years, winter precipitation in the Sierra Nevada and Rocky Mountains has been falling more and more in the form of rain rather than snow. The Southwest desert depends on the snowpack to continue to deliver water late into summer. With more rain and less snow, the summers will be getting drier.

Environmental Disasters

Nature generally stands to lose the most from global warming. Plant life and wildlife don't have air conditioning. Generally, plant life evolves slowly with the climate. If the climate gets progressively warmer, then trees and plants will migrate out of our parks and move northward. The trouble is that roads and cities break up the natural migration paths. Plants and trees are more likely to die out along with the wildlife they support, rather than migrate.

Polar Bears

Polar bears are adapted for living and hunting on sea ice. But global warming is melting the Arctic sea ice at an alarming rate. According to the National Snow and Ice Data Center at the University of Colorado at Boulder, sea ice is declining at a rate of more than 23,000 square miles per year (or nearly 9 percent per decade).

Polar bears are literally drowning from global warming—unable to swim the increasingly longer distances between land and receding sea ice. They depend on the ice to hunt seals, their primary food. With ice forming later in the fall and breaking up sooner in the spring, the time period bears can forage for food is shrinking throughout the far north.

Starving Whales

Gray whales, which migrate from the Bering Sea to Baja California each year, have been dying off in large numbers. The population dropped from an estimated peak of 26,635 whales in 1998 to 17,414 whales in 2002—a decline of more than one third.

Observations of whales in 2007 show that they are getting skinnier again, and another crash in their numbers is feared. The 35- to 50-ton beasts spend their summer in the Bering Sea gorging on millions of amphipods—crustaceans that live in tubes on the shallow ocean floor.

Although the whales migrate all the way from the lagoons off Baja California to the Bering Sea to eat, the amphipods rely on their food being brought to them by ocean currents. But global warming may be altering ice patterns and ocean currents in ways that prevent plankton from making its way through the water column to the amphipods. The whales, which feed on these shrimp-sized creatures, could be the largest victims yet of global warming.

IPCC Predictions

So … what can we expect for the future? The Intergovernmental Panel on Climate Change (IPCC), a group of more than 2,000 scientists appointed by the United Nations and the World Meteorological Organization, concluded in its 2007 report that the climate will rise between 2.0 to 11.5°F (1.1 to 6.4°C) by the end of the twenty-first century. The panel is made up of the superstars of climate research—university department heads and leading government agency researchers from all over the world. And they've been studying this stuff since 1988.

Global Warnings

Global warming might increase forest pests in Florida and throughout North America, which could weaken the trees and make the forests ripe for fire.

If the temperature only rises 2°F, we might be able to live with it. If it rises 11.5°F (6.4°C), then that's going to cause major problems. That's as great a change as ended the last ice age. Only instead of happening over about 13,000 years, it will happen in fewer than 10 decades. Some effects are expected to become apparent in the next 30 years.

Sea Level Rise

The IPCC predicts that sea level will rise in the range of 4 inches to 3 feet (.09 to .88 meters) over the next century. The greatest impact will likely be in recently developed coastal areas like Florida and parts of the U.S. Gulf and Atlantic coasts. In Florida, sea level rises of only a few inches will advance the shoreline as much as 400 feet in low-lying areas. This will flood shoreline homes and hotels and erode the state's treasured beaches. Sea level rise may also damage the Everglades, offshore islands, and coral reefs. In short, much of what has made Florida so precious may be lost in a globally warmer world.

Going to Death Valley

A warmer world is likely to also be a wetter one with catastrophic storms and floods over much of the globe. On the other hand, some areas like the plains and the deserts will get hotter and drier. Parts of the plains will turn into deserts.

In the Southwest desert where temperatures reach between 110 and 120°F (43 to 48.5°C) in the summers, temperatures could start climbing up to between 120 and 130°F (48.5 and 54°C). To get an idea of what that feels like, you have to go to Death Valley, California.

Death Valley is the hottest and driest place in the western hemisphere. On July 10, 1913, the mercury there reached 134°F—a world record for many years. When it gets above 115°F (46°C), the visitors center at Death Valley National Park wraps its door handles in Styrofoam so people won't burn their hands, and the maintenance staff keeps its tools in water so they won't burn theirs. People use gloves to drive their cars so the steering wheels won't burn them. Locals shut their water heaters off in late spring because the tap water is plenty hot enough.

Late July is the normal peak of yearly temperatures, with temperature readings regularly approaching 130°F (54°C). Around that time, there are lots of electrical outages where the air conditioning goes off for as much as four or five days at a stretch. Then, the park rangers sleep under wet sheets with wet hair. Or they sleep on picnic tables outside with the sprinklers on.

Park rangers describe Death Valley in July as being similar to roasting a turkey—when you've just opened the oven to a blast of hot air. The oven door is stuck … and so are you.

The Costs

Fixing global warming is not going to be cheap. The Kyoto agreement, which would have required the United States to reduce its greenhouse emissions 7 percent below 1990 levels, was only supposed to be a first step. Entirely stopping climate change will require 50 to 60 percent cuts. This is likely to be hugely expensive, requiring major reductions in the use of gasoline, fuel oil, natural gas, and coal.

Doing something now about global warming is like taking out insurance. We take out insurance on our cars and our homes in the event of major catastrophes. We don't expect disasters, but we want to be prepared.

Maybe it's time to contact an agent.

The Least You Need to Know

- ◆ Greenhouse gases (like carbon dioxide) keep us from freezing to death. We've increased its concentration in the atmosphere by burning fossil fuels.

- ◆ Scientists have studied tree rings, ice cores, and marine sediments to get a picture of our climate in the past and help them predict the future.

- ◆ Glacial ice is melting all over the world as a result of the current rate of global warming.

Part 6

The Solution

Okay, now that I've tackled the basics, discussed the politics, and laid out the pros and cons, it's time to address the problem. This isn't going to hurt. It might save you some money. And it might save the world some misery.

I'll start by looking around the house and discussing all the ways we can save energy. Some are simple fixes that won't cost anything. Others might require some investment, but you'll be saving on your electric and heating bills in the long run.

Then I'll invite you to take a look at your car. You don't have to go out and sell your eight-cylinder sedan today, but think about your options for future purchases. Do you really need that eight-passenger SUV, or could you and the earth live better and healthier on a little less horsepower? And what about planes? Is travel really such a good deal?

Finally, I'll take a look at some energy alternatives. With gas reserves on the wane, it's prudent to investigate. We can make our fossil fuels cleaner right now and make the cars that use those fuels a lot more efficient. Plus, there are some exciting new possibilities in solar, wind, and fuel cell technologies. It's time to make some changes.

"I want a representative who'll subsidize public transportation."

21

It Starts with the Individual

In This Chapter

- It's all about saving energy
- Conserving energy in the home and office
- Recycle and reuse
- How we get our electricity

Reversing global warming starts with the individual. It's a two-stage process. First you have to make a decision that it's a problem worth addressing. Then you have to make a commitment to do something about it. Most of the greenhouse gases in the atmosphere come from the burning of fossil fuels. You and I do this by using energy at home, in the office, and in our cars. I'll address the issue of cars in the following chapter. In this chapter, I'll talk about how we can cut back on energy usage where we live and where we work. I'll also talk about how we get our energy and what we can do to influence a cleaner generation of electricity. Cutting back on energy usage cuts back on the amount of carbon one releases into the environment. It today's terminology, it's called reducing your *carbon footprint*. Most everything in this chapter will help you reduce your carbon footprint.

def•i•ni•tion

Your **carbon footprint** refers to the amount of CO_2 your activities—driving a car, turning on the lights, running a refrigerator, and so on—give off into the environment.

In and Around the House

If you're sitting in your easy chair reading this book in a room with all the lights on, miffed about the draft coming through your closed window, thinking about checking that old refrigerator for something to eat, and looking forward to taking a nice, long, hot shower—have I got some news for you. You might just be a walking, talking, global warming machine. But don't worry. I've got some great stuff here for you to consider for reducing the amount of energy you consume.

The Windows

You're feeling a draft coming from the window, but it's closed. Don't worry. You're not losing your marbles—you're just losing precious heat through those flimsy portals of yours. Windows are typically the weakest points between your living room and the outdoors. The cheapest way to stop those drafts from running up your heating bill is to reduce the heat loss through the cracks and crevices around the window casings. Caulk around the edges of those drafty frames. Do it both on the inside and outside of your windows. Re-putty loose frames. Put some weather-stripping around the windows that are opened most often and around your doors. You've just gone a long way toward stopping heat loss through your windows.

Global Warnings

The amount of energy that leaks through windows in American homes each year is equal to the energy produced by the annual flow of oil through the Alaskan pipeline.

Now, look at the glass. If it's single-pane, it's not doing much more than blocking the wind. Double-pane windows provide an extra layer as well as some dead space between the panes of glass that can increase insulation dramatically. Storm windows add that extra layer as well. Most hardware stores will sell you plastic barriers that you can use to achieve the same effect during the winter months. They'll cut heat loss through single-pane windows by 25 to 40 percent. Curtains and shades are better than nothing but aren't as good as double-pane windows.

Insulation

Insulation can turn a drafty old house into a cozy home that is much cheaper to heat or cool. The first place you'll want to consider insulating is the ceiling or attic. It's where the majority of heat is lost or gained. The next most efficient place is the walls. After that, it's beneath the flooring. Insulation comes in a variety of forms, including loose-fill cellulose (made from recycled newspapers), fiberglass, foam board, insulated panels, and sprayed-on cellulose. The difference in the warmth of my home the day before I insulated the attic and the day after was dramatic.

> **Climatoids**
>
> Compact fluorescent bulbs screwed into ordinary sockets can provide incandescent light color, last up to 12 times longer, and use one fourth of the electricity. Start by replacing your most frequently used lights. Each incandescent bulb that you replace will prevent 1,000 to 2,000 pounds (453 to 906 kilograms) of carbon dioxide emissions over the life of the bulb.

The Lights

Lighting uses about one fourth of all the electricity consumed in the United States. That includes exterior and public lighting. You can cut your in-home lighting bill by 90 percent if you use the most efficient sources—including a little sun. The incandescent light bulb is the most wasteful kind of light you can use, both at home and in the office. About 90 percent of the energy it consumes is given off as heat, and only 10 percent is converted to light. Fluorescent lighting typically yields three to four times as much light for the same amount of energy.

Fluorescent lights are better than they used to be. The newer bulbs are much closer to natural sunlight than their incandescent counterparts. And new ballasts don't hum like the older ones. Fluorescent lights cost more than incandescent, but fluorescents can last up to 12 times as long. When you add the longevity of the fluorescent bulb to the savings you'll get on your electric bill, they end up being about half as expensive as an incandescent bulb.

The Refrigerator

Refrigerators and freezers consume about one sixth of all the energy used in a typical American home, but a lot of that can be saved. First, if you've got an old refrigerator, get rid of it. My family saved almost $25 a month on our electric bill when we traded

Global Warnings

The costliest refrigerator you have is that extra one in the garage that's keeping that six pack cold. Pull the plug.

in our old refrigerator for a new one. That was enough of a savings to cover the payment on the new one. Advances in technology have made the modern-day refrigerator about 60 percent more efficient than it was 20 years ago.

In general, the simpler refrigerators are the most energy efficient. Automatic ice makers typically increase energy use by 14 to 20 percent. Water dispensers waste energy as well. Check the "Energy Guide" labels that appear on the new models and compare the one you want with the others. Getting the one that uses the least energy doesn't mean your food is going to spoil. It means your bills will be lower and the atmosphere will be cleaner.

Mark packages and containers in your refrigerator so you can find what you are looking for quickly. Standing there scratching your head with the refrigerator door open is a waste of energy. And don't set the refrigerator thermostat any lower than you need.

The Heater

Think about getting a programmable thermostat that lets you set the heat for different parts of the house, and reduce it when you're not there or are fast asleep. Setting the thermostat back 10°F (5.5°C) at night can save 10 percent on your heating bill. You can get yourself a real nice blanket with those kinds of savings. Or buy yourself a sweater and keep the heat lower during the day. Just turning the thermostat down one degree cooler than normal will reduce carbon dioxide emissions from electrical generating facilities by 1,157 pounds (525 kilograms) a year.

Climatoids

A home energy audit, which most utility companies will provide free of charge, will help you identify ways you are wasting energy around your home and what you can do about them. The average family can save 1,000 lbs. of carbon dioxide emissions by following an energy auditor's advice—while saving money on family energy bills as well.

Air Conditioning

The same goes for air conditioning. Set your thermostat a little higher and you'll save energy. Don't turn on the air conditioning because you're too lazy to take off your sweatshirt. Try a ceiling fan instead of air conditioning. It's possible to cool your home without using a lot of energy. Install awnings on south-facing windows where there is no roof overhang. Hang white window shades or blinds, which can reflect as much as 40 to 50 percent of the midday heat.

The Water Heater

Heating your water is the second biggest use of home energy after heating and cooling. Heating water costs the typical family about $160 to $390 a year. If you use electricity to heat your water, then you are paying about twice as much as if you used natural gas or propane. Whether it's gas or electric, consider wrapping your water heater in an insulation blanket made especially for water heaters and you'll lose a lot less heat to the air.

Another way to lower your water-heating bill is to use less hot water. Energy-efficient shower heads can cut back on your water usage by half. They work by aerating the water or breaking it up into many tiny streams. You'll still get a good shower. You can use aerators on your kitchen and bathroom faucets, save energy, and still get the job done. Retrofitting one shower head and two faucets will reduce your annual carbon dioxide emissions by 580 to 3,200 pounds (261 to 1468 kilograms).

Climatoids _____

Turn your hot water down to 120°F (48.5°C) and your hot water costs will drop as much as 50 percent.

The Dishwasher

The dishwasher uses almost as much energy as a clothes dryer or freezer. Most modern machines have a booster to get the water temperature up to 140°F (60°C) to cut grease and kill germs. This isn't that bad, because it allows you to turn your hot water heater down to 120°F (48.5°C). The most efficient dishwashers on the market currently use about five gallons of hot water per load, and that's plenty. (Sailors aboard combat ships get two gallons to clean their entire bodies.) Consider using the light wash or energy-saving cycles when you can. Use the least amount of energy to get the job done, not the most. Choose air drying over heat drying; it'll make a difference. And scrape your dishes off, don't rinse them.

If you are going to rinse your dishes off, why not just wash them by hand? We do around our house. We had a dishwasher, but we got rid of it. We found that it didn't take any longer to wash the dishes than it took to load the dishwasher. My mother, however, disagrees. But my electric bill is lower than hers.

Clothes Washer and Dryer

A recent study at Cambridge University's Institute of Manufacturing found that 60 percent of the energy associated with a piece of clothing was spent in washing or drying it. Wash your clothes in warm water rather than hot. You'll save energy, and your clothes will last longer. Save up your dirty clothes and wash them all at once. Remember, newer machines use less energy than older ones. The least energy-consumptive way to dry your clothes is to hang them on the line out in the sun.

The Stove

Cooking appliances are some of the biggest energy consumers in the home. If you've got the air conditioning on, then the hot stove will make it work harder. Gas is your best bet. The fuel is used directly to cook the food. With electric, the fuel is brought to a power plant, converted to electricity, and then shipped long distances over power lines. It requires three or four units of fuel to produce one unit of electricity.

Consider a microwave for certain cooking tasks. They use up to two thirds less electricity than conventional electric ovens. Toaster ovens, slow cookers, and pressure cookers also save substantial energy costs in comparison with conventional ovens.

Electronics

Whether you are at home or in the office, keep your electronics off when not in use. Leaving your computer on doesn't make it run better; it only wastes energy. The same goes for your TV and stereo. If you want to leave a light on when you're not there to scare away prowlers, put it on a timer. It will look more natural (that is, it will scare more prowlers) and save energy as well.

When buying new electronics, look for the Energy Star label that will tell you the energy efficiency ratings. Energy Star was introduced by the U.S. Environmental Protection Agency (EPA) in 1992 as a voluntary labeling program designed to identify and promote energy-efficient products, which would reduce carbon dioxide emissions. Energy Star has expanded to include new homes, residential heating and cooling equipment, major appliances, office equipment, lighting, and consumer electronics. By using Energy Star appliances at home, consumers can reduce their utility bill by as much as 30 percent.

You Are What You Eat

You can also save energy and reduce greenhouse gas emissions by thinking about the food you eat. Those grapes from Chile may taste great in January, but think about the fuel that was wasted getting them to your door. Buy local when you can. Farmers markets are great places to pick up fresh fruits and vegetables in season. Enjoy your meals in comfort knowing that a minimum of fossil fuels was used to get them to your door.

Skip the steak. According to the U.N.'s Food and Agriculture Organization, the international meat industry generates roughly 18 percent of the world's greenhouse gas emissions. That's more than transportation. Think about becoming a vegetarian—at least, part of the time.

In the Garden

Not only do house and garden plants give off oxygen and absorb airborne toxins, they're good for our emotional health as well. Diane Relf, Ph.D., founder of the People Plant Council—a consortium of university horticultural therapists—says that just looking at plants and gardens can reduce blood pressure. In one study, patients who had a view of nature got out of the hospital faster, needed fewer painkillers, and even bothered their nurses less. Here are some hints to make your garden more environmentally kosher.

Pesticides

Pesticides for the most part are petrochemicals—petroleum products. They utilize fossil fuels and emit carbon dioxide in all stages of their production. They're also poison, and birds, butterflies, pets, and small children are sensitive to them. Yet the National Audubon Society reports that homeowners use three times more pesticides per acre than the average farmer.

In general, try the least-toxic product first. If that doesn't work, try something a little stronger. But utilize the least amount of toxins to get the job done. For ants, fill a sprayer with a few tablespoons of soap, add water, and spray the ants. It kills them and cleans the chemical trails they leave to attract other ants. For mice or rats, use mechanical traps rather than poisons. Bait the traps with peanut butter, which is more attractive to them than cheese.

Spray your plants with insecticidal soaps, which kill insects but are not harmful to you or other animals and will not emit greenhouse gases. Soaps for both insects and funguses are made from plant materials and are available at most nurseries. Or simply wash bugs off with a stream of water. Do it early in the day so the moisture will dry and funguses won't flourish. Flowering plants like daisies will attract predatory bugs and help contain pests. Or, buy your own predatory bugs—like ladybugs—and release them in the garden. They're available at many nurseries. Keeping your garden in a natural balance will produce much healthier plants and flowers.

Use native plants where you can. They need less water and are more resistant to pests and disease, so you'll need less pesticide. Look for nurseries that specialize in them. Visit your local arboretum and scout up some good native possibilities. It's a little more trouble to find them but a lot easier taking care of them.

Compost

Consider starting a compost pile in the corner of your yard. The United States spends up to $5 billion a year on fossil fuel-derived fertilizers that release chemicals into the ground and cause the release of nitrous oxide, another greenhouse gas. Instead of putting the garbage down the drain, put it on the compost pile and you will be helping keep our rivers cleaner. Pile grass clippings, weeds, and leaves on it and the garbage man won't have to burn energy to haul those items to the dump. And your plants will love the dark, loamy mulch that comes out of it. There are classes available in many communities to show you how to get started. Don't put meat, fish, or any animal products on your compost pile and the pile won't attract insects.

Putting a little manure or organic fertilizer on your compost will get it going and decaying. Water it once in a while. Also, take a shovel or spade and turn it over every month or so and it will break down faster. I've got a compost pile in my yard, and I use the mulch I get for my roses. They are the best in the neighborhood.

Recycling

Reuse what you have, and recycle things you can't use. By reusing things, we save energy and resources needed for the future. We also reduce the greenhouse gas emissions and pollution that occur from producing something new.

Producing new paper, glass, and metal products from recyclable materials saves 70 to 90 percent of the energy, pollution, and greenhouse emissions that would result from using virgin materials. You can save a good-sized tree by recycling a stack of newspapers only 4 inches (10 centimeters) high. Support recycling efforts in your community, and tell your local officials that you think it is important. Why throw away perfectly good resources in our landfills? We don't have enough space left. The typical American discards 1,300 pounds (590 kilograms) of trash each year, and we're running out of places to bury it. Every year, more than 200 million trees are cut down to make U.S. newspapers—the majority of which are tossed into the trash. Americans discard enough aluminum cans annually to rebuild the entire U.S. commercial airline fleet four times over.

Recycle Clothing

Buying vintage clothing is a lot more environmentally sound and climate friendly than buying new stuff. Buying a shirt the second time around means that all the carbon that was emitted manufacturing and shipping the new one isn't emitted again. If you are tired of your old clothes, have a "closet swap." Invite some friends over and tell them to bring a few items that are nice but may not fit or they're just plain tired of. Then trade. It's a great way to renew your wardrobe while being environmentally friendly.

Climatoids

Have a green wedding. If your relatives are traveling a long way to your special event, offset the carbon emissions by donating to renewable-energy projects. Or plant some trees. Make sure your wedding feast is as local as possible. Buy local wines and foods so they don't have to be shipped in along with your relatives.

The Price of Coal-Fired Energy

Electrical generation from fossil fuels, particularly coal, is the single largest source of greenhouse gas emissions and air pollution in the United States. Coal-fired power plants emit more than 31 percent of the nation's carbon dioxide. But as I mentioned in Chapter 19, the burning of fossil fuels and the production of carbon dioxide comes with a generous portion of fellow pollutants as well. Coal-fired electrical generating facilities account for more than 60 percent of the nation's sulfur dioxide emissions and 23 percent of its nitrogen oxide emissions.

Acid Rain

Acid rain is formed when sulfur dioxide and nitrogen oxides react with water and oxygen in the atmosphere to form acidic compounds. The deep South and the Midwest are home to the largest number of coal-fired power plants, but they may escape the worst acid rain damage. That's because damage drifts eastward and falls on the eastern coasts of the United States and Canada.

Acid rain ruins lakes. Insect populations that surround acidic lakes are much less diverse than those around normal lakes. This affects fish populations. Scientists have identified 200 lakes in the Adirondacks and 200 in Ontario where all fish—including trout, walleye, and bass—have disappeared because acid rain made the waters too acidic for them to survive. The same goes for frogs, toads, salamanders, and other amphibians. The acidification of our natural waters hinders the reproduction of a number of waterfowl, including ring-necked ducks, mergansers, and loons. It also affects river otters.

Acid rain eats away our buildings and monuments, too. Limestone and marble, the stones that form many of the buildings and monuments in Washington, D.C., are especially vulnerable to acid precipitation because these monuments were built with materials that contain calcium carbonate, which dissolves easily in acid. The U.S. National Park Service reports that the buildings affected include the U.S. Capitol, the Smithsonian Institution, the Arlington Memorial Bridge, the Lincoln and Jefferson Memorials, and the Washington Monument.

Other Pollutants

Coal-fired power plants are the largest source of mercury emissions in the United States, accounting for an estimated 32 percent of emissions. Mercury in the air eventually ends up in lakes, rivers, and coastal waters, where it is a deadly poison. Predator fish such as salmon, lake trout, and walleye may have concentrations of mercury more than a million times higher than the surrounding waters.

Coal-fired plants also emit ground-level ozone—a powerful respiratory irritant that is particularly dangerous for the elderly, children, and people who have asthma or other respiratory diseases. Ozone triggers more than 6 million asthma attacks each year.

Researchers at Carnegie Mellon University in Pittsburgh studied the health effects of pollution from fossil fuels on death rates in four international cities including Sao Paulo, Brazil; Mexico City; Santiago, Chile; and New York. They found that the burning of coal and gasoline is causing people to die prematurely from asthma, heart disease, and lung disorders. They estimate that more people are killed by air pollution than by auto accidents.

Old Power Plants Kill

The passage of the Clean Air Act in the United States has cleaned up many power plants and reduced sulfur dioxide emissions. However, there is a loophole in the Clean Air Act that allows older, dirtier power plants to emit more pollutants than newer, more efficient plants. A number of plants in the South and Midwest have used this loophole to make significant upgrades to their facilities without addressing the environmental issues. The Sierra Club estimates that as many as 600 existing power plants between 30 and 50 years old are 10 times as dirty as new power plants built today.

Congress set up this loophole because they assumed that the older plants would be replaced by newer facilities, but power companies are using the loophole to extend the life of plants way beyond their years. It needs to be stopped. Let your representative know how you feel.

Alternatives to Coal

English lawmakers switched their country's electric power plants from coal to natural gas and as a result had no trouble meeting Kyoto standards. The same could happen in North America. There are new, efficient technologies for natural gas power plants that require less energy to run and would save money over the lifetime of the facility. Natural gas is inexpensive and has the lowest carbon content per unit of energy of all the fossil fuels.

Abundant, clean, and reliable energy sources such as wind, biomass, and solar power are catching on in North America. Sacramento, California, announced a $26 million program to use solar power to run more than 3,000 homes—the largest solar project in the nation. Mid-American Utilities has opened the world's largest wind power plant in Buena Vista County, Iowa. The government needs to encourage more efforts like these through tax cuts, rebates, and other incentives to reduce our reliance on fossil fuels. Call, write, or e-mail your legislators and ask what they are doing about these issues. If everyone who read this book did just one of these things just once, it would make an enormous difference.

The Least You Need to Know

- ◆ The production of electricity is one of the major sources of global warming in North America. Cut down on greenhouse gas emissions by being more energy efficient at home.

- Newer appliances are much more energy efficient than older ones. Look for Energy Star labels on new products to determine which are the most efficient.

- You can reduce carbon dioxide emissions by recycling around the home and by composting in the garden.

- Electrical generation from fossil fuels, particularly coal, is the single largest source of greenhouse gas emissions and air pollution in the United States.

- Switching to natural gas and alternative energy solutions to generate our electricity could do a lot to clean up the air.

Chapter 22

What About Transportation?

In This Chapter

◆ Raising fuel economy standards

◆ The hidden costs of SUVs

◆ Hybrid and electric cars

◆ Tips on saving gas

◆ What about planes?

We like to get around, but cars and planes are large contributors to global warming. Let's look at our motor vehicles first—not the fronts, but the backs where the tailpipes are. Automobile and truck emissions are one of the largest sources of carbon dioxide in America today. In fact, every gallon of gas burned creates 28 pounds (12.2 kilograms) of carbon dioxide in the atmosphere—19 pounds from the tailpipe and 9 from refining and transporting the fuel. The average car creates about 70 tons of carbon dioxide over its lifetime; the average sport-utility vehicle (SUV) around 100 tons. Raising fuel efficiency standards is the largest single step we can take to curb global warming.

But that's not our only transportation problem. Planes do their share of changing our climate. Let's discuss some options.

What Would Jesus Drive?

When Congress failed to increase fuel efficiency standards in 2003, a broad coalition of religious leaders got together and formed the National Religious Partnership for the Environment. It included the U.S. Conference of Catholic Bishops, the National Council of Churches U.S.A., the Coalition on the Environment and Jewish Life, and the Evangelical Environmental Network. The organization immediately began to lobby Detroit, Michigan, auto executives asking for improvements in fuel economy. They claimed they had a biblical mandate to be good stewards of God's creations. They also said they felt a responsibility to the poor who were especially harmed by pollution.

Climatoids

Current U.S. Corporate Average Fuel Economy (CAFE) standards require new cars to average 27.5 miles per gallon (mpg) while light trucks, SUVs, and vans must average 22.2 mpg. By contrast, the current standard in Japan is 30.3 mpg and in Europe 33 mpg. There are no breaks for SUVs. U.S. environmental organizations contend that fuel economy could be raised to between 35 and 40 mpg over the next 10 years using current technology—without creating a burden on the automobile industry, which contributes 4 percent of the country's gross domestic product.

One of the smaller groups within the religious partnership, the Evangelical Environmental Network, decided it would take things a step further. The group purchased TV spots and ads in national magazines that posed the question, "What Would Jesus Drive?" For that matter, what would any of the great prophets—Moses, Mohammed, Confucius, Buddha—drive if they were running around in the flesh today? Would they pull up in a Humvee, spouting fumes and sucking up gas? Probably not. So … why should we?

Top Five Global Polluters

America's cars and light trucks alone produce 20 percent of the U.S. carbon dioxide pollution. The pollution is not only a threat to our atmosphere but also a threat to our health. According to the American Lung Association, nearly 117 million Americans live in areas where the air is unhealthy to breathe. Light trucks and SUVs magnify the health risk by spewing up to three times more smog-forming pollution than regular cars.

U.S. autos alone emit more carbon dioxide than the total output (autos, industry, agriculture—everything) of all but four countries. What this means is that if we were to rate the world's top five emitters of carbon dioxide, the list would look like this:

1. United States

2. China

3. Russia

4. Japan

5. U.S. automobiles

The more gas you guzzle, the more carbon dioxide you are going to emit. When you compare the total tonnage of carbon dioxide produced over an average car's 125,000-mile lifetime, you get a clearer picture of the problem. According to the Sierra Club, a Ford Excursion, which gets 13 miles per gallon, emits 134 tons of carbon dioxide. A Ford Crown Victoria, which gets 21 mpg, emits 83 tons. A Ford Escape Hybrid, 33 mpg, emits 51 tons. A Toyota Prius at 55 mpg emits 32 tons. The Ford Excursion emits 4 times as much carbon dioxide in its lifetime as the Toyota Prius.

The SUV and Light-Truck Loophole

Back in 1975 when fuel efficiency minimums were first adopted under the Corporate Average Fuel Economy (CAFE) standards, the term "light truck" referred to a vehicle that was used to haul hay on the farm and lumber at construction sites. Lawmakers reacting to the petroleum crisis in 1973 and growing amounts of auto exhaust pollution passed this legislation, which set the makers of American sedans on a course toward lighter and more fuel-efficient cars. The CAFE standards, however, included exemptions for light trucks, which were then used by farmers and workers who needed more powerful autos for their jobs.

But what happened is that American automakers used the light truck definition to skirt the law. Back when the law was first passed, light trucks, SUVs, and minivans made up only 20 percent of the vehicles sold in America. Today, they make up more than half the vehicles sold. But these cars are more likely to haul lattés home from Starbucks than lumber from the yard. Although many are used as family cars, they are more often seen with single drivers than with multiple riders.

The hidden cost of SUVs is the price we pay with our natural resources. To keep the gas tanks of these vehicles full, oil companies seek to drill in new areas—including some of our nation's most sensitive wilderness habitats. These include the fragile coastlines of Florida and California, the lands around Yellowstone National Park, and the Arctic National Wildlife Refuge in Alaska.

The *Exxon Valdez* disaster off Alaska in 1989 and the breakup of the oil tanker *Prestige* off the coast of Spain in 2002 serve to remind us of the dangers of transporting petroleum products in and around our national and international treasures. Small spills and oil leaks, which never make the newspaper but are no less disastrous, happen all the time.

Global Warnings

Switching from driving an average car to a 13-mpg SUV for one year would, according to the Sierra Club, waste more energy than if you …

- ◆ Left your refrigerator open for 6 years
- ◆ Left your bathroom light burning for 30 years
- ◆ Left your color television on for 28 years

Auto industry advertising portrays SUVs as the ticket to the great outdoors. They are shown splashing through creeks, churning up mud, running along the beach, and climbing high, rocky mountains. In reality, most of them never see dirt unless they're accidentally tearing up the lawn next to the driveway.

SUV Safety

Although many people buy SUVs for safety purposes, the opposite is actually true. The height that makes the SUV driver feel above the traffic and safer is the Achilles' heel of the vehicle. When SUVs swerve to avoid an accident, they are four times more likely to roll over. Rollovers account for 22 percent of deaths in normal cars but 62 percent in SUVs.

Hot Debates

According to a government study in the 1990s, at least 2,000 traffic fatalities would not have occurred had their cars collided with another car rather than a light truck or SUV.

Because an SUV is taller, heavier, and more rigid, it's more than twice as likely to kill the other driver in an accident. That's because when an SUV and a car collide, the height difference, stiff frame, and greater mass of the SUV create a weapon that can be lethal to passengers. They're also lethal to pedestrians, bicyclists, and motorcycles. Braking standards for SUVs are weaker than for cars.

What's Your Mileage?

What can you do? You can start by figuring out what kind of gas mileage your car gets. It's not that difficult. The next time you go to the gas pump, write down the mileage. You have to fill the car up or you can't figure the mileage. Then, the next time you get gas, fill the car up again and note how many gallons you used. Also note how many miles you drove, either by subtracting the old odometer reading from the new or by using one of the alternate readings on the odometer. Divide the number of miles you drove by the gallons you used (what it took to fill it up again). The result is your mileage per gallon of gas. My wife's Honda Civic Hybrid, however, flashes its miles per gallon from the panel controls. But then, at an average of 50 mpg, the car has something to brag about.

Try it a few times. Note the difference in the mileage when you drive it around town and when you drive it on the highway. Pay attention to how your mileage goes down when you drive it over 60 mph (96 kph). If your car is in tune, you'll get better mileage and you'll be dumping less carbon dioxide into the atmosphere. You can increase the fuel efficiency of your car by keeping your car maintained and by watching your driving habits.

Auto Maintenance

A well-maintained car will get better mileage and emit less greenhouse gas. You'll spend less at the pump and the atmosphere will be cleaner. Here are some tips:

- **Tune that engine.** A finely tuned engine has more power, idles better, gets better mileage, and creates fewer emissions. Get your car tuned once a year. Follow the manufacturer's recommendations. Shop around for a good mechanic; the cheapest one isn't always the best one. Check your oxygen sensor regularly. Make sure the catalytic converter is working properly. Simply replacing the spark plugs can make a difference. A car that's out of tune, has a faulty or plugged catalytic converter, or a faulty oxygen sensor can reduce your mileage by as much as 40 percent. That means that you may be using 40 percent more gas than you need. That can mount up fast.

- **Check the air filter.** Have your mechanic check your air filter and replace it regularly. An air filter catches the dirt and other impurities before they enter your engine. That dirt can wreck your engine as well as your mileage. Replacing a dirty filter can increase your mileage up to 10 percent.

◆ **Keep those tires inflated.** This may be the easiest one of all, but it's important. Making sure your tires are properly inflated will increase your mileage, make your car handle better, and make it safer to drive. It's a good idea to use your own tire gauge to check your tires. The tire gauges on the air hoses at gas stations are notoriously inaccurate. Get a gauge at an auto parts store and check your tires before and after you fill them up.

◆ **Get the right oil.** Use the manufacturer's recommended grade of oil. Using a 10W-30 oil in an engine designed to use 5W-30 can lower your mileage. Keep your oil clean. Manufacturers tend to give you a range of mileage recommendations for oil changes. Consider using the lower mileage to keep your engine running smoothly.

Global Warnings _____

Gas mileage decreases rapidly as you drive more than 60 mph (96 kph). Each 5 mph (8 kph) you drive over 60 mph (96 kph) is like paying an extra 10 cents per gallon of gas. Not only is high-speed driving risky to you and the vehicles around you, it's expensive.

◆ **Watch that lead foot.** If you like to speed, accelerate rapidly, slam on the brakes, weave in and out of traffic … guess what? You're wasting gas. Not only are you a lot more apt to get into accidents, but your mileage will also suffer in the process.

◆ **Don't idle.** When you idle your car, you are getting 0 mpg. The bigger the car, the more gas you waste when you idle. Don't leave your engine running unnecessarily. Driving cautiously for the first few minutes is a better way to warm up your car than to sit there idling.

◆ **Just cruise.** If you've got cruise control, use it. Driving at a constant speed saves gas. Use your overdrive gear on the highway. It will reduce the revolutions per minute (rpms) of your engine, reduce engine wear, and save gas.

Climatoids _____

Consider carpooling, even if it's just one day a week. Many urban areas allow vehicles with multiple passengers to ride in special high-occupancy lanes. If every commuter car carried just one more passenger, we'd save 32 million gallons of gasoline every working day.

◆ **Combine those trips.** Got errands? Well, do them all at once rather than separately. If you combine your stops, you'll save time and money. Several short trips taken from a cold start will use up to twice as much fuel as a longer, multi-purpose trip covering the same distance but with a warm engine. Your car is much more efficient when it's warmed up. Cold starts are expensive. Keep them to a minimum.

◆ **Try telecommuting.** If you can go to and from work during the off hours, you'll avoid traffic, get there more quickly, and use less fuel. Tell your boss it's a win-win situation. Ask him or her if you couldn't telecommute from your home one or two days a week. That will save a lot of mileage. If you own more than one vehicle, drive the one that gets the best mileage. Save the truck or SUV for times when you really need it.

◆ **Take the train or walk.**   Take the train or other public transit. You'll not only arrive more refreshed, but you can take the time to read the paper or do some work during your commute. Or consider walking or riding a bicycle. Just trimming five miles off your normal driving pattern can eliminate 2,000 pounds of carbon dioxide in the atmosphere each year.

◆ **Use roof racks wisely.**   Roof racks are great if you've got a small car and need to take a trip. They'll give you more cargo space, and you can enjoy the economy and efficiency of the smaller car in town. But be advised that loads carried up top increase the aerodynamic drag of the car and decrease your fuel efficiency. If you have the option, stick your stuff in the trunk; don't carry it up top.

Choose a More Efficient Vehicle

I recently purchased a four-wheel-drive pickup truck. It takes me places out in nature I need to get to that a passenger car won't go. But I got a four-cylinder rather than a six. I previously owned a six-cylinder car and frankly couldn't imagine getting a four-cylinder. The last four-cylinder I had was a Volkswagen Bus. I owned it for several years and had to replace the engine twice.

But … I wanted to be ecological, so I went ahead and got a four-cylinder. I expected it to be underpowered, but it wasn't. Cars are far better engineered these days than back in the 1970s when I was driving my bus around. My truck gets good mileage, is economical, and has plenty of power. I go up hills just fine. I might downshift, but I don't have to pull over to let the other cars pass me. If I enter the highway and need to get up to speed, there's plenty of it. Even loaded up with traveling and camping gear, it runs fine.

I essentially have a light truck, the one we've been complaining about, but it *is* a four-cylinder. My cousin and my sister-in-law both have Toyota *hybrids*. They claim they have no problem on the highway. My sister-in-law brags how she drove from Los Angeles to San Francisco (425 miles—680 km) on 10

def•i•ni•tion

A **hybrid** vehicle is one that utilizes both an electrical motor and a gas-powered engine that work in tandem to substantially decrease fuel consumption.

gallons (37 liters) of gas! My wife just got a Honda Civic Hybrid, and now we mostly drive her car. It's amazing how many trips she can take without stopping for gas.

Hybrid Cars

Hybrid cars are the first alternative-powered cars on the market that seem to be catching on. They save gas by combining an electric motor with a gasoline-powered one. More than 150,000 were sold in the first few years after their introduction. Global cumulative sales of Toyota hybrid vehicles have topped the one million mark as of May 31, 2007, according to the company. Toyota and Honda were the first on the scene with hybrids. But now, a number of American companies afraid of ceding another fast-moving market to the Japanese are coming out with their own. Hybrid SUVs might soon counter a little of the SUV's bad press.

The electrical motor in the hybrid gets its energy from batteries that are charged by the gasoline engine when it is running or by collecting energy from the car's brakes when they are applied. In the Honda Civic and the Insight, an electric motor assists when the car is climbing a hill or accelerating sharply. In the Toyota Prius, the electric motor takes over at low speeds. In both cars, the gas engine shuts off and the electric one takes over when the car is stopped.

The hybrid's rise has been encouraged by rising gas prices, clamoring from environmentalists, and by stricter environmental regulations in a number of states (most notably California). That state mandated that a certain percentage of these cars be made available if car makers wanted to do business there. California is one of the biggest car markets in the world and also suffers some of the nation's worst pollution.

Climatoids

Even the U.S. Army has been jumping on the hybrid bandwagon. The Pentagon is developing hybrid electric engines for the replacement for the Humvee to extend the vehicle's range and increase its survival on the battlefield. In Iraq, the war is fought throughout the country, and supply convoys are always at risk. Hybrid vehicles could reduce the number of convoys. The military also appreciates the fact that in electric mode the car runs silently, making it great for reconnaissance missions.

Sport-utility hybrids are in the works that will soon start getting 40 mpg (16.6 kpl) or better compared to recent averages closer to 20 mpg (8.4 kpl). Although these cars currently cost about $5,000 more than other models without the hybrid function, that cost is expected to drop as production increases. Even now, if you factor in state and federal tax incentives to buy the car with the gas savings over the life of the car, you'll come out even. Ford, Mazda, Mercury, Lexus, Chevy, and GM all have their own hybrids and more are planned. GM even has a hybrid city bus.

Plane Travel

The view from my campsite atop a 13,000-foot ridge in the Andes Mountains, just inside Manu National Park, Peru, is at once breathtaking and troublesome. Out over the Amazon Basin, giant cotton tufts filled with vapor are rising up, amassing their assault on the cloud forests below my perch. For the next couple weeks I will be rambling beneath the forests' dense canopy draped with mosses and orchids and witnessing a sampling of the 1,666 species of birds found here in the Andes at various times of the year.

Visiting natural places like the Andes is one of the reasons why I became a journalist. But I can't help but ask myself: Isn't my wanderlust—and that of my readers—part of the climate change problem about which I'm often reporting? Ecotourism, while educational and enriching, also contributes to the increasing emissions that are causing climate change. "One of the worst environmental acts a person can do is buying an airplane ticket," says Christina Cavalier, director of training and education for the International Ecotourism Society. The average coast-to-coast flight can expend 1.23 tons of carbon dioxide per person.

Even without the flying, travel can be a significant source of greenhouse gas emissions. Lots of things we do—driving, staying in hotels, eating out, even buying souvenirs—have carbon tolls. Now, a growing number of organizations are on hand to help travelers calculate their carbon emissions and offset them by planting trees or supporting renewable energy sources that can, in effect, make travel *carbon neutral*—a term that was added to the *New Oxford American Dictionary* in 2007. It's essentially the same as carbon trading—a concept most often associated with the Kyoto Treaty—and now it's available to individual consumers at the click of a mouse.

def•i•ni•tion

You can make your travel **carbon neutral** by offsetting the carbon emissions made from planes, trains, and cars with purchases from organizations that will plant trees or invest in alternative energy in amounts that offset your travel distances. "Carbon neutral" is a term that was added to the *New Oxford American Dictionary* in 2007.

Zeroing Out Your Carbon Debt

There are many calculator websites to choose from, with prices and carbon estimates varying widely. I first log on to the website for Native Energy (www.nativeenergy. com), the company that Al Gore selected to zero out the carbon emissions created from making *An Inconvenient Truth*. Native Energy has an option on its home page that lets you calculate carbon emissions from flying, driving, taking buses, and staying in hotels.

The mileage printed on my plane ticket shows I covered 10,505 miles from Los Angeles, California, to Lima, Peru. Add to that my estimated distance of 125 miles for the train ride to Machu Picchu and 150 miles for the bus ride to Manu National Park, and my carbon debt adds up to a total of 4.41 tons. Rounding up, I can pay $12 per ton for five tons, or $60, to offset my entire trip while helping to fund the construction of alternate sources of energy, including windmills and methane generators.

When I try to replicate these steps on Travelocity, which now offers a Go Zero program allowing customers to "zero out" carbon emissions generated by their flights, hotels, or driving, I am only given the option to make my trip carbon-neutral if I buy a ticket—making it impossible to comparison shop.

The Conservation Fund's website (www.conservationfund.org) is more user-friendly. Although it asks me to input estimates about my annual travel, there's nothing stopping me from entering the mileage for just one trip. When I type in my air miles for Peru, it spits out a toll of 2.31 tons of carbon dioxide (slightly less than half the Native Energy estimate). For only $14.24, I can plant two trees—which over the next 70 years will take approximately 2.31 tons of carbon from the air and lock it into the tree. The Conservation Fund gives no estimate for cabs, trains, and hotels, but if I round out my total to 3 tons I will still only pay $17. The donation goes to plant native trees in once-forested areas (most of them protected lands in the Mississippi Valley).

Better Just to Stay Home?

Now, we come to the obvious question: "Is it better just to stay home?" "Of course," says Lovins L. Hunter, president and founder of Natural Capitalism, Inc. in Eldorado Springs, Colorado, and a professor of business at Presidio School of Management in San Francisco, California. "But that's not realistic. We want to drink from streams as yet unseen. I want to see the footprint of a snow leopard in Afghanistan." Lovins points to small experiments in air travel with ethanol and hydrogen. Says Lovins,

"Offsets are better than nothing, but we need to come up with alternatives to carbon-based fuels."

Despite its problems, travel can be important for conservation. Travel is crucial to host communities and an important economic activity in developing countries. Plus, travel is a learning tool. It gives the traveler an awareness of the world and the environment that you won't get from staying home watching television.

The Least You Need to Know

- Raising the fuel efficiency standards of our vehicles would be the most effective thing we could do to stop global warming.

- Light trucks and SUVs get around fuel efficiency standards established in the United States in the 1970s because the law considers them work vehicles. Little work is done in these vehicles; they need to be held to national standards.

- Hybrid cars are the most fuel-efficient promise presently on the market, but driving sensibly and keeping your car in tune will help our air.

- Cheap gas might be a luxury the environment can't afford.

Chapter 23

The Rainforest Is on Fire

In This Chapter

- Treasure in the tropical rainforest
- Involving the natives
- Ecotourism
- The jungle needs help

The tropical rainforests of the world are a unique treasure. Rainforests are found around the world, mostly near the equator. They support an incredible array of wildlife. Scientists estimate that anywhere from 50 to 90 percent of all animal species are found in tropical rainforests, and many of these species haven't yet been discovered. But rainforests are under attack from farmers, ranchers, miners, oil companies, and developers trying to clear the land. Satellite images of the Amazon jungle show more than 8,000 fires burning in a single day.

As the trees are cut and burned, they release a tremendous amount of carbon dioxide into the atmosphere. The destruction of the forest and the tilling of the land produce about 25 percent of the greenhouse gases released by humans into the atmosphere. The tropical rainforest needs some friends. Let's hold a concert.

Concert in the Jungle

I am sitting in the ruins of the Plaza Mayor, the ceremonial gathering point of the 1,400-year-old Mayan City of Tikal in Guatemala, waiting for the concert to begin. It's 5:30 A.M., and the sunlight is just starting to filter through the thick jungle canopy that envelops the ruins.

The symphony begins with the flutelike songs of the melodious black birds, followed by a chorus of high caws from the Aztec parakeets and the red lored parrots. The tempo picks up with the hysterical humanlike cackles of the laughing falcons as the lineated woodpeckers pound out the backbeat on tall swaying palms. The crescendo continues to rise until daybreak, when the howler monkeys join their avian brothers and sisters with a cacophony of lionlike roars that can be heard for miles.

Tikal is to Guatemala what Banff National Park is to Canada or Yellowstone National Park is to the United States. Protected since 1955, the wildlife within the park's boundaries is both plentiful and oblivious to human intruders. Tikal is located in the subtropical forest of the state of El Petén in northern Guatemala. Within the boundaries of the park, the forest is thick and virgin—almost completely undisturbed.

Global Warnings

Scientists believe that the collapse of the Mayans might have been due to climate changes as well as the mismanagement of their forests. As forests disappear, so does wildlife and local plant species. The collapse of the Mayans might be one of history's first ecological catastrophes.

In populated areas just outside the park, however, the forest makes up only a quarter of the terrain—and very little of that is undisturbed. Guatemala is a microcosm of what is happening in other tropical third-world nations in Latin America, Africa, and southeast Asia. Virgin rainforest originally covered 4.2 million square miles (11 million square kilometers) of the tropics, but now only 380,000 square miles (1 million square kilometers) remain.

Lessons from the Maya

Tikal was a thriving center of Mayan commerce. At its height between 700 and 800 C.E., this citadel once held a population of more than 90,000 Mayan Indians with more than 4,000 structures—including plazas, temples, palaces, ball courts, and pyramids. A number of these monumental structures have been excavated, although most are still buried beneath the dense green jungle. The collapse of the Mayas, scientists speculate, is attributable to both drought and deforestation—the latter of which led to a collapse of the natural food chain that surrounded them.

The Rainforest as a Sink

The rainforest is a vital component in the take-up and release of carbon dioxide in the atmosphere. Trees and other plants absorb carbon dioxide while growing, and they release carbon dioxide when they decay. They also release carbon dioxide when they are burned. The reason why the tropical forest is so important to the atmosphere is because of its density.

This is why it's such a good idea to save it. The density of the forest creates an enormous sponge that can forgive a lot of our industrial sins by sucking up carbon dioxide from the atmosphere. The Kyoto agreement allowed industrial nations to receive credits toward their reductions of greenhouse gases by replanting tropical forest.

> **Climatoids** _____
>
> A few countries, such as Spain, have demonstrated that temperate forests can be restored to the land even after centuries of abuse. Spain has reforested nearly 5 million acres (2 million hectares) since the end of World War II.

Regrowing temperate forest is likewise a good idea. Spain has done a good job of that. But temperate forest can't match the density of plant life and wildlife that is in the tropical rainforest—nor can the temperate forest match the rainforest's carbon absorption ability. And saving the tropical rainforest now is a lot easier than replanting it later, because the growth of tropical trees is much slower than those in more northern climates.

Slash and Burn

Tikal is located in the northern lowlands of Guatemala in the Province of Petén, which is typical of agriculturally developed tropical jungles. Although the jungle (or tropical rainforest, as it is variously referred to) is a vital and rich biological system, the soils are as good as sterile. "You scrape away the leaf litter," says Dave Whiticre, head of the Maya Project that is studying the raptor community in the Petén for the Peregrine Fund, "and you'll see there is a thick layer of leaf litter and essentially no humus. You get down to mineral soil almost immediately."

The leaf litter is rapidly decomposed by the incredible mass of jungle insects. Leaf-cutter ants cut paths 6 or 7 inches (15 to 18 centimeters) through the leaf litter in Tikal. And the heavy rains leach nitrogen and potassium out of the soil. The nutrients in the tropics are essentially held within the tree, so when the forest is destroyed, the land is poor for agriculture.

Global Warnings _____

Deforestation in the tropics can change the weather hundreds of kilometers away. Scientists have found that reduction of the lowland forest affects the amount of moisture that rises with the clouds to the higher mountain jungles. The resultant reduced cloud cover is affecting the high jungles in Costa Rica, extending the dry season and diminishing amphibian populations.

On the road from the Flores airport in Petén to Tikal, Whiticre and I stop the car at a section of patchwork jungle that exposes the series of successions that typically occur with the slash-and-burn method of agriculture. Here, farmers take a tract of forest and cut it down during the dry season. They let the wood dry for a month or so and then torch it before the rainy season. Thus, the nutrients in the trees are released in the form of ash. The farmer tills the soil and plants his crop just before the rains.

During the first season, the crop is good, the insect pests and weeds are at a minimum, and the soil is rich—but with each successive crop, the soil gets poorer, the insects and the weeds build up, and the output grows weaker. In some areas, farmers can get 5 to 10 crop rotations—but in most areas, after two or three rotations the field is abandoned and the farmer moves on to cut another section of the jungle.

Black Earth Can Help

Along the Amazon River, archaeologists of the Central Amazon Project have discovered that early indigenous societies had an answer to slash-and-burn agriculture. Some people call it slash and char. The Central Amazon Project is an international consortium of archaeologists that attracts the best and the brightest in the profession from Brazil, Chile, Argentina, Colombia, Panama, the United States, and the United Kingdom. These scientists have discovered that before Christopher Columbus, indigenous people on the peninsula between the Amazon and the Rio Negro rivers had towns and cities with buildings, defensive structures, moats, agriculture, roads, and trade. Although they cleared some forests, they used lower-temperature fires—perhaps covering the trees with damp grasses that turned the felled timber into charcoal.

This charcoal, or *black earth*, was then mixed into the soil. What the charcoal did for the soil was to attract and retain essential plant nutrients like nitrogen, phosphorous, and potassium—the same nutrients that are in those fertilizers you can buy at your local nursery. With slash and burn, those nutrients are washed away after several seasons—but with slash and char, the black earth that remains can stay fertile for more than 1,000 years. Soils built by these early societies, some 2,000 years ago, are still fertile today.

def•i•ni•tion

Black earth is a type of soil created by early indigenous societies in the Amazon that used charcoal from slowly smoldered trees. When mixed into the soil, it held back plant nutrients that were usually lost with the incessant tropical rains. The use of black earth by current settlers, scientists feel, could eliminate the need to burn down more tropical forests.

Farmers on the Central Amazon Project peninsula all want to plant in black earth. Farmers from the project, which is near Manaus all the way to the mouth of the Amazon, look for black earth left by early societies. Jim Petersen, an archeologist from the University of Vermont who was murdered in the Amazon in August 2005, truly believed that black earth could greatly reduce deforestation in the Amazon by eliminating the need for settlers to continue their cycle of burning. The increased fertility of the soils could also go a long way toward stopping tropical third-world hunger. And it would stop global warming from the burning of tropical forests—the cause of 20 percent of climate change. Archeologists such as Brazilian Eduardo Neves (from the University of São Paulo) and American William Woods (from the University of Kansas) have been trying to carry on Petersen's work to spread the message about black earth.

Archeologists working in black earth in the Amazon.

(Photo courtesy of Michael Tennesen)

Conservation Approaches

Conserving the tropical rainforest requires a variety of approaches. Conservationists need to provide economic alternatives to local populations so natives can make a living doing something besides cutting down the trees. Conservationists need to provide education so locals can learn the value of the rainforest and the species contained. And conservationists need to work with international corporations to make sure business doesn't contribute to further *biodiversity* loss.

def•i•ni•tion _____

Biodiversity describes the number of species in a given environment. It is shorthand for "biological diversity."

Alternative Rainforest Products

Forest communities create enterprises from oil, nuts, fruit, plants, fibers, and other natural materials sustainably harvested from trees and plants—thereby creating an incentive to keep the forest standing. In Petén, a number of locals harvest chicle from the sapodilla tree, an evergreen that flourishes in the rainforest. Guatemalans harvest the sap without cutting down the trees. The Mayan Indians chewed chicle more than 1,000 years ago. Chewing gum made with chicle was developed in the 1860s.

Much of the coffee and cocoa produced in this region is grown on plantations where the forest was cut back to encourage the growth of the beans. But many farmers still produce shade-grown coffee and cocoa, which has become popular in the United States in organic markets. Starbucks even has shade-grown coffee. By choosing shade-grown (organic) coffee or cocoa, North Americans can be sure that all of the trees weren't cut down to make the coffee or cocoa beans grow faster. These trees are vital to wildlife and the uptake of carbon dioxide.

Historically, all coffee was cultivated in the shade beneath a forested canopy. But in the early 1970s, fostered by grants from the U.S. government, farmers in several countries began growing varieties of coffee that thrive in direct sunlight. The goal: to increase production. But without the natural protection provided by a surrounding forest, sun-grown coffee usually requires heavy doses of pesticides and chemical fertilizers. Researchers at Conservation International found that in some places, growers use about a pound of toxic chemicals for every pound of sun-grown coffee they produce—making it one of the most chemically intensive crops on the planet.

You can encourage the saving of the rainforest by the kind of coffee you drink. Shade-grown coffee (most organic coffee is shade-grown) is the one to ask for. Ninety percent of the coffee purchased in America is sun-grown, which means that the rainforest was cut down to maximize growth. By asking for shade-grown coffee or organic coffee the next time you're at your local coffee shop or grocery store, you'll be making sure that major portions of the rainforest remain intact.

Ecotourism

Ecotourism is another way we can help the tropical rainforest. The term ecotourism began as little more than advertising copy for high-priced adventure travel in the 1970s and 1980s. But since 1990, it has become a buzzword in both the travel industry and the environmental movement. It really has two meanings. In the crowded natural parks and wilderness areas in the United States and Canada, ecotourism refers to minimizing visitor impact on our parks and wild places. Elsewhere it refers to a growing part of the travel industry that takes people to exotic places where the environment is factored into the vacation.

def•i•ni•tion

Ecotourism is environmentally friendly tourism. In North America, it refers mostly to minimizing visitor impact. In third-world nations, it more often refers to supporting local economic alternatives to depleting natural habitats.

At Home

Ecotourists in the Everglades put canoes into a special section of the park, known as the "Nightmare," where motor boats are not allowed. The Nightmare is a narrow section of the swamp, overhung by mangroves, where visitors can see egrets, herons, roseate spoonbills, and osprey. Alligators occasionally jump into the water as you pass.

Managing tourism's impact in natural environments sometimes requires methods that some argue defeat the purpose of pristine areas. One example is Isle Royale National Park, which is 90 percent wilderness. In this forested wilderness where the "hand of man" is not supposed to show, the National Park Service has built boardwalks in many of the boggy areas. Park rangers point out that when people walk through a boggy area, the mosses break down underfoot—creating muddy trails that gradually widen with continual traffic. With a wooden path, however, the damage is contained.

The tent cottages at Maho Bay Camps in the Virgin Islands National Park on St. John have won numerous awards for blending into their tropical background and for the camps' environmentally friendly approach. The tent cottages, built on 16-square-foot platforms, are made of translucent fabric on wooden frames with screened terraces to catch the cooling trade winds. The facilities utilize solar and wind energy, composting toilets, and energy-efficient appliances. Old newspaper is the only protective wrapping used at the Maho Bay Store, and the camps' bar gives a 25-cent discount if guests will reuse the same mug on refills of beer. To stay at such an ecohotel, check out www. greenhotels.com, which lists such establishments from Philadelphia to Hawaii.

Abroad

The trend toward ecotourism in international parks and wildernesses is a bit different. Ecotourism in foreign countries has centered more on support for local communities so that locals will have a reason not to exploit timber, wildlife, and other natural resources in protected areas.

Vendors displaying beautiful textiles and wooden carvings are located along the road up to Tikal National Park. By hiring a local guide and by purchasing native crafts, you are giving them an alternate source of income to cutting and burning the forest to raise corn. Conservationists believe that most Latin American parks are not endangered by too much tourism. With the funding problems that most protected areas have in Latin America, what is most needed is a heck of a lot more tourists.

Getting the Natives Involved

A few years ago, I traveled to Vilcabamba in the Peruvian rainforest with an international team of biologists sponsored by Conservation International. They were trying to assess the species diversity of this high mountainous region between the Andes and the Amazon for consideration as a park. I got to witness how the biologists worked to involve locals in their work.

Early one morning, I accompanied Tom Schulenberg, an ornithologist at Chicago's Field Museum and leader of the expedition. With us was Lawrence Lopez, a Peruvian ornithologist whom Schulenberg was trying to train. The biologists worked around the edge of the bog that centered our study site, training both binoculars and microphones on the surrounding jungle. According to Schulenberg, you can hear approximately three to four times more birds than you can see, and he was anxious to teach this ability to Lopez. All American biologists on the expedition had Peruvian counterparts who they were trying to train.

The population of Peru is gradually becoming more environmentally aware. Monica Roma got her Master's degree in ecology at the University of Missouri and remembers how in 1995 when she returned to Peru, ecology was the fashion. Five years earlier, no one even knew the word. "But when I got back, people were protecting the dolphins, recycling their garbage, and talking about saving the rainforest."

Our Way

International conservationist groups encourage the World Bank and the Inter-American Development Bank to emphasize environmental needs when loans are made. Those development loans channel millions of dollars into poor countries for road projects, dam building, the creation of export-agricultural plantations, and other useful efforts. But poorly planned projects funded by multilateral lenders have resulted in human colonization of rainforests and in agricultural projects that weren't always in the best interests of the rainforest.

Dalia Amor, a Ph.D. student at the Nicholas School at Duke University, has been working with Latin American biologists and conservationists to protect the jaguar in Central America. Farms and roads have cut into jaguar habitat in Central America, although the animal needs large tracts of contiguous forest to maintain healthy populations.

Toward this end, Amor has been working on a project called the Mesoamerican Biological Corridor, which will join enough protected areas to allow the jaguar to travel from Panama to Mexico. The Inter-American Development Bank has invested in this goal, but at the same time it has invested in a $4 billion plan that entails the construction of more than 332 dams and 4,000 square miles of roads in the same areas. Says Amor, "This is crazy; it will only negate the efforts behind the Biological Corridor."

Biologist Dalia Amor with jaguar.

(Photo courtesy of Michael Tennesen)

Wilderness Parks

One of Conservation International's principal strategies focuses on major tropical wilderness areas. The three principal areas they promote are Amazonia in South America, the Congo Forest in Africa, and New Guinea in Indonesia. These areas are the largest remaining tracts of pristine tropical forest on Earth and are more than 70 percent intact. Tropical wilderness areas are among the last places where indigenous people can maintain traditional lifestyles. And they are of crucial importance to climate regulation.

Roads

Part of the urgency that biologists gave to the expedition into Vilcabamba was that oil giants Shell, Mobil, and Chevron had won concessions from the Peruvian government to explore the valleys east of Vilcabamba, extending to Bolivia. If oil or natural gas was to be found, a pipeline might be built over the Andes.

Things like pipelines, oil wells, and commercial logging make biologists and conservationists nervous. All of these activities introduce roads that provide access to the deep forest. That access brings illegal hunting, indigenous farms, and population growth.

Global Warnings

Conservationists are often less afraid of the oil exploration, mining, and commercial logging operations in the tropical rainforest than they are of the roads these enterprises bring. After the oil wells are gone, the roads they leave behind provide access to formerly inaccessible regions of the forest, bringing poaching, slash-and-burn agriculture, ranches, villages, and the eventual demise of the tropical rainforest.

Hotspots

Besides promoting wilderness, Conservation International also promotes what it calls hotspots. Hotspot is a term coined to identify areas with large numbers of species that are concentrated in a relatively small space. If we can save these areas, we will be able to preserve a maximum number of species for the limited amount of conservation dollars available.

Conservation International lists some 25 areas as hotspots. Those include the tropical Andes, Madagascar, South Africa's Cape Floristic region, the Antilles, the Philippines,

Brazil's Atlantic forest region, the Polynesian and Micronesian island complex including Hawaii, southwestern Australia, the Mediterranean region, the Guinean forests of west Africa, and others.

The 25 hotspots identified by Conservation International contain 44 percent of all plant species and 35 percent of all terrestrial vertebrate species in only 1.4 percent of the planet's land area.

Global Warnings

Hotspots have the highest concentrations of unique biodiversity on the planet. They are also the places at the greatest risk of destruction. The need for conservation in these regions is urgent to prevent a wave of species extinctions.

Diversity

Vilcabamba is in the middle of the tropical Andes and is incredibly diverse. Many biologists believe that the tropical Andes are the most biologically diverse of all the remaining forests in the world. The Vilcabamba Range, roughly the size of New Hampshire, is an uninhabited cloud forest in the midst of this towering wilderness. Cut off from the nearby landmasses by the deep valleys of the Apurimac and Urubamba rivers, it rises like an island in a sea of jungle and is as isolated as an island in the ocean.

Animals in such remote landscapes are isolated, too. Separated from the rest of the world, many populations have evolved alone and have been out of touch with each other over the eons. The result: pockets of diversity, often teeming with species never before described by scientists.

Climatoids

Biologists investigating the Vilcabamba Range, an uninhabited cloud forest roughly the size of New Hampshire between the Andes and the Amazon in eastern Peru, estimate that the area has at least 650 species of birds. That number is roughly equivalent to the total number of species that breed in the United States and Canada.

Their study sites demonstrated the incredible diversity of the land. Although its elevation was the upper limit for monkeys, there were white-faced capuchin monkeys, spider monkeys, and night monkeys. There were signs of puma, dwarf deer, and bear.

The mammalogists on the team collected a weasel, a rare short-tailed opossum that was previously represented by only two Peruvian specimens, and a rare bamboo rat that had hands like a monkey—a first for both the Peruvian and Smithsonian collections. They also encountered 16 species of reptiles and amphibians with as many as 12 being new to science. The biologists catalogued 142 species of birds. Schulenberg estimated that Vilcabamba had at least 650 species of birds.

One day I accompanied Brad Boyle, a Canadian botanist who specialized in South American plants, and his Peruvian counterparts, two young Peruvian biologists. We traversed the bog to the forest beyond, then climbed to a rise where they laid out a 50-yard line and took specimens of the plant life within two to four yards of either side of the line.

That day, their attention was directed upward at the epiphytes—bromeliads, orchids, ferns, mosses, and other plants—that crowded the limbs of the trees above. The enormous presence of these epiphytes is one of the factors that distinguish cloud forest from other types of jungle. Boyle pointed to a cluster of treetops in one area and declared that there were more species in that cluster than in most northern deciduous forests.

Cloud forest animals often specialize on individual plants or groups of plants. Some flowers have long, curved tubes that only cycle-billed or sword-billed hummingbirds can pollinate. But there are also cheaters, like flower piercers, that have hooked bills like a can opener that can make little holes at the base of the flower, and there are cheater bees and other smaller hummingbirds that use those holes to get at the flower's nectar.

Boyle expected to find many new species in such an isolated forest but said describing new species was a long and arduous task. "It's not like, 'Eureka! I just found a new species'; it's more like, 'Wow, I've just been through every known species in the herbarium and it doesn't match anything.'" Still, he held up a tiny delicate orchid and dared to declare, "I'll bet you a case of beer that's never been described before."

Specialists

Plants and animals are distributed differentially throughout the globe. Some, like humans and gray rats, are widespread and inhabit most regions while others have very restricted distributions and may be found only on a single island, in a single river or lake, or on a single mountaintop. We refer to these highly restricted organisms as *endemic* species because they are unique to a specific region. Many of the species that occurred in the Vilcabamba range were endemic—found nowhere else in the world.

def•i•ni•tion

Endemic species of plants or animals are those that are found only in restricted areas—a single island, a single river or lake, a single mountain range, or a single mountaintop.

Ants Won't Stop Progress

The areas where endemic species flourish are often difficult, far removed, and treacherous. Vilcabamba was one of these difficult places. When I exited the helicopter, my leather boots sank six to eight inches into a bog. They stayed wet the entire two weeks I was there. The earth, saturated with water, was covered with a deep, spongy matrix of moss. Here and there were sinkholes that could swallow your leg to the thigh.

The team cleared paths with machetes. The branches that remained after the machetes were sharp and acted as natural booby traps if you walked too rapidly. The trails were like obstacle courses with stumps and tangled roots. I learned to walk on all fours. One of the Smithsonian biologists, Louise Emmons, slipped on the trail and fell into one of these cut branches. Although there was no x-ray machine in camp, she felt she'd cracked her rib. The worst part was that she'd been nursing a bronchial infection for a number of days and now couldn't cough.

In the evening, I thought about taking a bath, but Brad Boyle warned me, "The army ants are swarming down by the river. Be careful. You don't want to get one up your pant leg." I decided to wash my clothes instead. Wool socks took two days to dry. My cotton T-shirts never dried.

At night I thought of the beauty of the cloud forest—the orchids, the mosses, the ferns—but I also thought of the hardships. I wondered if Vilcabamba was tough enough to save itself. I proposed this to Louise Emmons, but she disagreed. "Too many times we've assumed that places were indestructible or too remote, and we come back 10 years later and they're gone."

If the world wants to have the tropical rainforest in its future, we need to start saving now.

The Least You Need to Know

- The destruction of the forest and the tilling of the land creates about 25 percent of the greenhouse gasses released by humans.

- Tropical rainforests are the greatest storehouse of species diversity and density on the planet. The rainforest is also a vital component in the take-up of CO_2 from the atmosphere.

- Agriculture, mining, oil exploration, and the roads they bring all pose serious threats to the continued existence of the rainforest.

- We need to provide alternative methods of commerce to native populations if we are going to keep them from cutting the rainforest down.

Chapter 24

Running Out of Gas

In This Chapter

◆ Bustling futures in clean fuel technologies

◆ Solar, wind, and other alternative sources

◆ Clean nuclear energy

◆ Everyone's wild about hydrogen

Although electricity was used in the early 1800s, it wasn't until Thomas Edison opened the first electricity-generating plant in London in 1882 that the energy age really got going. Cities got electricity first. Rural areas in the United States didn't get theirs until the 1930s, but it's been growing ever since. The U.S. population grew 79 percent from 1949 to 1997, but electric usage grew by more than 1,100 percent.

And we've been looking everywhere for coal, oil, and natural gas ever since. About 80 percent of the oil produced today flows from fields that were found before 1973, and the great majority of them are declining. New exploration tools haven't improved the oil discovery rate, even though today's tools and techniques are much improved from earlier decades.

Oil: What's Left?

U.S. oil production peaked in the 1970s and has been going down ever since. Many scientists believe the world's production will follow the same production peak and decline in the next 10 years. We will experience severe shortages of natural gas first, followed by oil, followed by coal.

The problem is that our cleanest energy is natural gas—what we're trying to convert to. The dirtiest is coal. China has the greatest reserves of coal, but if it ends up using it to fuel accelerating development, it could neutralize Western efforts to reduce pollution and greenhouse gases.

But ... don't despair. A dose of global warming may bring on depression, but the antidote could be the best thing that ever happened. Opportunity is knocking. Fossil fuels have a diminishing future. Alternative energies have an unlimited one. It will be hard, but we can do it. We went to the moon; this is a piece of cake. Besides, there's money to be made.

At the end of 2005, world-proven crude oil reserves stood at 1,153,962 million barrels, of which 904,255 million barrels—or 78.4 percent—was in OPEC member countries. The founding members include Iran, Iraq, Kuwait, Saudi Arabia, and Venezuela. According to OPEC estimates, at the rate of production in 2005, OPEC's oil reserves are sufficient to last more than 80 years while non-OPEC oil producers' reserves might last fewer than 30 years. The worldwide demand for oil is rising, and OPEC is expected to be an increasingly important source of that oil. Does that explain some of our current problems?

What Else Can We Use?

Although it may take time to develop a full suite of energy alternatives, there are things we can do right now to clean up the energy we've got.

Coal

Coal produced 75 percent of our energy by the end of World War I, but then oil moved in and began to fuel our trucks, cars, and locomotives. Oil was cleaner and easier to move around. Natural gas muscled in and took over the cooking and heating market. But coal moved back into dominance in the late 1970s, when it became the principal fuel for electric utilities.

The trouble is, coal is about 70 percent carbon and produces more carbon dioxide than oil or natural gas. It also emits more sulfur dioxide and nitrogen oxides. In the United States, the recent trend has been for electric utilities to move away from coal toward natural gas. Similar changes in Britain have radically reduced its greenhouse emissions.

It's possible to make coal cleaner. There are techniques that can remove up to 90 percent of the sulfur and ash. Those techniques include using microwaves, microbes, and enzymes. Coal gas has promise as a cleaner-burning fuel. Mixing finely ground coal with oil or water to produce liquid can reduce pollutants.

Climatoids

Residences account for 21 percent of U.S. CO_2 emissions. More than two thirds (68 percent) of residential emissions come from the consumption of electricity, and most of those emissions (80 percent) come from the burning of coal at coal-fired power plants. Carbon dioxide emissions from the generation of electricity account for 39 percent of total U.S. energy-related CO_2 emissions.

Oil

We've been using oil for 5,000 years to light our lamps, lubricate our machinery, seal our ships, and make roads (and even weapons). Ancient Arabs made a kind of napalm out of it. But it was the invention of the internal combustion engine that really turned us into an oil economy.

The trouble is, burning oil creates pollution. It can be made cleaner, though. Reformulated gasoline, which is meant to reduce smog, is increasingly used across North America—particularly in California and the Northeast.

Natural Gas

Natural gas is mostly methane with ethane, propane, and a few other hydrocarbons mixed in. Methane, as you recall, is one of the greenhouse gases—although it has a shorter life in the atmosphere than carbon dioxide. When natural gas is burned, it's cleaner than any of the other fossil fuels. This is why the Clean Air Act Amendments of 1990 encouraged increased use of natural gas because of its low combustion emissions.

The Hoover Dam.

(Photo courtesy of Michael Tennesen)

Hydroelectric Power

Hydroelectric power is both cheap and clean. It's also renewable, because the water that drives the generators comes back each rainy season. Hydroelectric power normally comes from dams, which create reservoirs. As the water goes through the dam, the force is used to turn generators—creating electricity. Although hydroelectric power is gentle to the atmosphere, it's tough on the land, fish, aquatic species, Native Americans, and fishermen who utilize the river.

def•i•ni•tion

Hydroelectric power or electricity is created by the force of water driving a generator.

Dams can stop fish migrations, wreck spawning grounds, and slow the downstream flow of water, which impacts wildlife. Although fish ladders allow most of the fish to circumvent the dams, the environmental impacts of these structures are considered serious enough that few new projects are in the works in North America.

Diamond Valley Lake near Hemet, California, is an exception to that rule. The lake, about 4 miles long and 2 miles wide, is Southern California's largest freshwater reservoir. No stream, creek, or river was dammed to build it, so its environmental impact was minimal. Instead, three natural rock dams joined a corral of stately mountains to form the enormous reservoir, which was then filled by an aqueduct.

Nuclear Power

From a global warming point of view, nuclear power has certain advantages. It emits few greenhouse gases. France obtains 80 percent of its electricity from nuclear power. Other countries that rely on its use have fewer problems meeting Kyoto emission agreements than the United States. But alas, nuclear power has lots of bad points as well—like the Chernobyl and Three Mile Island accidents. North America and Europe have stopped building new plants and have turned an eye toward other nations who use nuclear fuels out of worry that nuclear bombs might be their intent.

Fission

Current nuclear power plants utilize fission to make energy. Heavy elements such as uranium are bombarded with neutrons, which break the uranium atom and release more neutrons. The process becomes self-sustaining, and each time an atom is broken it releases energy. That energy is used to heat water, which drives a turbine that produces power. Each reactor core contains several thousand fuel assemblies composed of sealed fuel rods filled with uranium oxide pellets. However, over time these rods lose their heating ability and the assemblies have to be refueled. What to do with the spent fuel is part of the reason why nuclear power is on the wane in most western nations.

Fusion

But then, there is fusion. Fusion is what the sun runs on. Fusion occurs on the sun when hydrogen combines to form helium or other heavier elements. It's a hot combination, releasing about a million times more heat than burning oil. The problem is, hydrogen nuclei normally repulse each other—so you need a lot of energy to bring them together. Right now, it takes more energy to cause fusion than you get out of the process. We've learned how to induce fusion in the hydrogen bomb, but you can't plug your stereo into it. But there are projects in the works, particularly in Japan, that could get this going.

The Promise of Solar Energy

The promise of solar power is more immediate. Solar power has come a long way since its early days. Vast areas in countries around the world that don't have power lines have fueled the growth. Solar energy comes in a variety of forms: solar water heating, passive solar heating and cooling, photovoltaic technology, and solar thermal technology.

Solar water heating uses a solar collector to absorb the sun's energy and then heat the water in a solar tank. A black 50-gallon drum filled with water set up on the roof of the house is the preferred solar water heater in Mexico's Baja Desert. Fancier devices work in higher latitudes and can deliver a surprising amount of hot water with no carbon dioxide or other pollution created.

Passive solar heating works with a building's design to heat the house in the winter and cool it in the summer. Adobe homes in New Mexico lean on early Native American designs where windows face south to capture the sun's low rays in the winter but are shaded by an awning to prevent those rays from entering the house in summer, when the sun hangs higher in the sky. Landscaping that uses deciduous trees, which provide shade in the summer but lose their leaves in winter to allow in sun, follow the same principle.

Photovoltaic Cells

Photovoltaic (PV) cells hold the greatest promise for solar energy. They take in light from one side of the cell and send electricity out the other. It's as simple as that. Yet, developing and improving these cells has been a monumental task. PV cells are created from various forms of silicon as well as film materials—some thinner than a human hair.

PV cells, however, create direct current, and because U.S. circuits run on alternating current, an inverter is needed. Most systems need switches to tie the PV system to the power grid so that the grid can act as a backup. In some areas, it's possible to run excess energy into the power grid and get credit for the contribution. A PV solar system also needs batteries to store the energy. This provides the ability to charge up the batteries during the day and to power the lights at night. Fluorescent lighting helps these charges last longer.

Climatoids

To supply the United States with its current energy budget would take an array of photovoltaic cells covering 10,000 square miles (26,000 square kilometers). A massive, but not insurmountable, scale-up is required to provide current U.S. energy requirements and to gear up for the 2050s, when energy requirements may triple.

Solar energy can be made more efficient by concentrating sunlight into the cell. It's better used in areas that are mostly cloud-free. This requires a tracking device to align the sun with lenses or reflectors that concentrate sunlight into PV cells.

The Oak Ridge National Laboratory has developed a hybrid solar lighting system that uses a roof-mounted solar collector to concentrate visible sunlight into a bundle of plastic optical fibers. The optical fibers penetrate the roof and distribute the sunlight to multiple "hybrid" luminaires within the building. The hybrid luminaires blend the natural light with artificial light (of variable intensity) to maintain a constant level of room lighting. One collector powers about eight fluorescent hybrid light fixtures, which can illuminate about 1,000 square feet. The system is connected to standard power sources to ensure that disruptions in service do not occur on cloudy days or at night.

Solar Thermal

Whereas photovoltaic cells process light into electricity directly, solar thermal energy requires using the sun's rays to heat a fluid that is then used to power a steam generator. Some systems use parabolic dishes, which feature a mirror array to concentrate sunlight into a receiver that holds liquid that can be used to drive a generator.

Currently, solar thermal devices do everything from heating swimming pools to creating steam for electricity generation. The U.S. Department of Energy reports that as of 2005, some 225 solar collection firms have shipped a total of 98,481 units to the market.

Solar Power from Space

Some scientists think that space is the best place to collect solar energy and so have been casting about for ways to make that possible. In space, sunlight is available almost all the time with no clouds or atmospheric conditions to dilute it. In the 1970s, NASA studied the idea of using a solar cell array the size of Manhattan in stationary orbit over the earth. The satellite would gather the light and convert it to radio frequency energy. It would then beam that energy to Earth, where a receiver would capture it and convert the radio frequency back to electricity.

Japan has invested in such a system and is working to beam solar energy to developing nations a few degrees from the equator from a satellite in low equatorial orbit. A number of tropical nations including Papua New Guinea, Indonesia, Ecuador, and Colombia have agreed to participate in the experiment. Japan hopes to be beaming down energy by the second half of the century.

Hot Debates
Wind energy is cheap and clean, but it has some drawbacks. Although wind farms are generally located in unpopulated areas, local residents complain about the visual pollution these huge wind farms create. Also, migrating birds—particularly hawks and eagles—can get tangled in the turbines and perish. One solution to this problem has been to put wind turbines out to sea.

Wind Energy

From high mountain passes in California to the shores of the North Sea, tall farms of wind turbines stand upright against the blustery force of the winds to capture their force in turbine blades and convert it to electricity. Wind turbines around the world are now producing commercial quantities of electricity without emitting any greenhouse gases.

Wind energy is actually another form of solar energy because wind is driven by temperature and pressure differences on the surface of the planet, which are caused by the sun. Wind turbines—generally with two or three blades—collect the energy of the moving air to drive a generator that creates electricity. Wind turbines are set on towers in places where winds blow hard and steady. They're different than windmills used on farms to bring up water. These monsters can have blades up to 89 feet (27.1 meters) long.

The U.S. Department of Energy estimates that global winds could provide more than 15 times the world's energy consumption. There is enormous potential for greater use of wind energy in the United States. Midwest states such as South Dakota, North Dakota, Nebraska, and Kansas could become the Middle East of wind power.

In windswept northern Europe, wind turbine operators are constructing huge, full-scale marine wind parks in the North Sea. There are advantages to offshore windmills besides the lack of visual pollution. Offshore, operators never run out of wind. It blows 90 percent of the time. With electricity production accounting for 40 percent of the world's emissions, the large-scale deployment of renewable energies is essential for establishing truly low carbon economies. Wind energy is booming around the world, with an average annual growth of more than 28 percent in the last 10 years—particularly in Europe.

Geothermal Energy

Geothermal energy arises from the earth, usually in the form of volcanoes, hot springs, and steam vents. But there's more down there than that, including decaying substances and friction along fault lines. If we could harness all the energy in just the top six miles of the earth's crust, it would equal 50,000 times the energy of all the world's gas and oil reserves.

Steam-dominated sources supply the energy for most geothermal power plants in the United States. The Geysers near Morgan Springs, California, is one of them. Here, water descends to relatively shallow depths from 1,000 to 10,000 feet (300 to 3,000 meters), heats up, and returns as hot water or steam. Hot rock could also be used, with energy suppliers pumping the water down into the rock and utilizing the heat that returns to run electric generators.

Geothermal energy could eventually power some 25 million homes across the United States at a cost of $40 million per year, according to a Department of Energy-sponsored study. Led by Jefferson Tester, a professor at the Massachusetts Institute of Technology (MIT) and lead author, the report argues that unlocking the heat stored in Earth's crust could improve American energy security while reducing emissions of greenhouse gases blamed for global warming.

Climatoids _____

Although geothermal-created electricity has a high price tag attached to its installation, using geysers or hot spring water to heat homes or businesses is a viable alternative. Geothermal heat pumps can move the water from 5 to 50 feet below the structure up into the building to heat it in the winter and even cool it in the summer, because subsurface temperatures stay nearly constant year-round.

Biomass

We get biomass energy from organic materials. The most common biomass material is wood. More than 500 facilities use wood or wood waste to generate electricity. Some use waste materials left over from planting, pruning, and harvesting. We can also make liquid and gas fuels like ethanol and methanol, which could be mixed with existing fuels. These would burn relatively cleaner than fossil fuels (although they are not as clean as geothermal, wind, or solar).

If we had a fuel crop, it could serve as an interim fuel until cleaner options were more available. One possibility is to use the plant residue from sugar cane, left after milling. Sugar cane was the basis for the world's first renewable biofuel program in Brazil. Corn is presently used to create ethanol in the United States.

Municipal solid waste is composed of 60 percent organic materials and can be used to generate energy. Landfills produce methane gas, which can be captured and used to fire electric generators. Suppliers can use yeast to ferment biomass and convert it to alcohol, which can be used for cars and trucks.

Biodiesel

What would you say to a vehicle fuel that is domestically produced, renewable, and produces dramatically fewer greenhouse gases than petroleum-based fuels? These are exactly the characteristics of biodiesel.

Rudolph Diesel, in a search for a more efficient alternative to the steam engine, invented the diesel engine in 1900. He then sought to help farmers by producing an engine that could run on fuel "grown" by those same farmers. The first diesel engine ran on peanut oil. Way back in 1912, Diesel predicted today's revival: "The use of vegetable oils for engine fuels may seem insignificant today. But such oils may become in the course of time as important as petroleum and the coal tar products of the present time."

Biodiesel can be burned in a conventional diesel engine with virtually no modification. The two major caveats are that biodiesel is a stronger solvent than conventional diesel fuel and may require the replacement of rubber fuel-line components with synthetic rubber; second, biodiesel may require additional fuel heaters in the fuel system in order to run in colder climates.

The Ocean as an Energy Source

There is a lot of power in the ocean from thermal, tidal, and wave-generating forces. There have been a number of experiments to capture the sea's reservoir of energy. One Japanese experiment in Hawaii and on the Pacific island of Nauru tried to harness the ocean's thermal energy by using warmer waters at the surface and cooler waters underneath to run a steam turbine. Others have tried to harness the energy of waves, using the surging, the up-and-down, and even the rocking motion of the waves—but none of these experiments have hit pay dirt yet.

The Canadians and the French have successfully harnessed the energy in the tides. Rising tides enter a reservoir controlled by gates that close at high tide. After the tide goes out, the reservoir releases the water to drive turbines much as if a river were passing through a dam.

> **Climatoids**
>
> Blue Energy of Canada is developing a technology that acts as an underwater wind farm. Turbines are mounted on structures that are set out on the ocean floor in places with strong tides or currents. Water flows through the structures, driving the turbines.

Fuel Cells

Fuel cells are not new. NASA developed the technology as a lightweight power source for U.S. space vehicles, including the space shuttle. Fuel cells work somewhat like a battery. In a battery, energy is stored on electrodes. "The battery runs until the charge runs out, and you have to recharge it," says John D. O'Sullivan, program manager for fuel cell technology at the Electronic Power Research Institute in Menlo Park, California, "but fuel cells provide continuous energy because you just keep adding fuel."

Fuel cells work by allowing oxygen to react with natural gas, methanol, or hydrogen to produce electricity without combustion. In hydrogen fuel cells, water and heat are the only by-products.

Both Honda and Toyota have introduced a few fuel cell vehicles in California, and more could be on the way. Every major automaker and oil company has a hydrogen or fuel cell research effort under way. Supporters say they recognize that fossil fuels can't last forever. Environmentalists complain that the industry is simply trying to avoid more immediate steps to improve fuel efficiency of conventional automobiles. Although it may be a while, fuel cells have that look of the future.

One of the problems is the price. It costs about $4,000 to make a gas engine. It costs about $40,000 right now to make a fuel cell. Fuel cells have few moving parts, so maintenance would be low. They don't consume themselves like batteries, and they remain almost endlessly rechargeable. Producing a similar amount of electricity as one fuel cell from a power plant would dump 1,100 tons of carbon into the atmosphere.

Because hydrogen, the ideal fuel source for fuel cells, is so difficult to create and store, fuel cells that run on gasoline or other less environmentally friendly fuels may precede the pure hydrogen models. Such a vehicle could travel 80 miles on a single gallon. It would emit no pollutants except for CO_2. The increased mileage would cut heavily into that.

A Hydrogen Economy?

Hydrogen has enormous potential. It could just be the prophet that finally leads us out of this sea of fossil fuels. Even now, just adding 5 percent hydrogen to the gasoline-air mixture in internal combustion engines could reduce nitrogen oxide emissions by up to 40 percent. Hydrogen has the potential for use as an aircraft fuel. Within a few years, hydrogen could be blended with natural gas by electric utilities wishing to reduce emissions. Standalone hydrogen-powered fuel cells could provide power for remote and commercial applications.

The major problem with using hydrogen as an energy source is that it doesn't occur naturally in any usable from. Hydrogen almost always bonds with something else and requires energy to break the bond to produce pure hydrogen. This is the rub. It currently takes twice as much energy to create hydrogen than it takes to make electricity.

Climatoids

Engines may be converted to run on hydrogen by the use of a SmartPlug, a device developed by kNew Corporation. SmartPlug replaces spark plugs, combining the functions of fuel injection and spark ignition, and can operate on a number of different fuels (including hydrogen).

Scientists are experimenting with a number of different methods for producing hydrogen. Most involve trying to break good old H_2O into hydrogen and oxygen. One method uses light focused on semiconductors submerged in water. Another uses dissolved metals as a catalyst to create water-splitting reactions. Some types of bacteria work to convert carbon monoxide and water to carbon dioxide and hydrogen. Scientists from the Woods Hole Oceanographic Institution recently boarded a submersible to search the hydrothermal vents at the bottom of the ocean for microbes that may one day be engineered to capture carbon dioxide and to strip hydrogen from methane.

Let's Get Started

So what are we waiting for? The challenge is out there. The dangers of just waiting around and doing nothing are becoming clearer every day. We've been dealing in oil for over a century. It's been good while it lasted, but it's turning on us. Following its path is increasingly dangerous.

It's time to start looking at other options.

The Least You Need to Know

- There are a lot of ways we can start cleaning up the fuels we use right now.

- Solar energy, including spaced-based solar systems, has enormous potential. Wind, geothermal, and biomass energy could fill in the gaps.

- Fuel cells run on hydrogen and may someday replace the internal combustion engine. If we can figure out a cheap way to create hydrogen, we might have global warming licked.

- Time's running out. Let's get going.

Glossary

acid rain   Rain, snow, sleet, hail, and fog, which because of the addition of pollutants becomes more acidic than normal.

aerobic   Living or occurring with oxygen.

albedo   The ability of a surface to reflect back the light that falls upon it.

anaerobic   Living or occurring without oxygen.

anemia   A disease in which body tissues are deprived of oxygen by a reduction in the number of red blood cells or inadequate amounts of an essential protein called hemoglobin.

auroras   The aurora borealis and the aurora australis, known as the northern and southern lights, are vivid nighttime displays of colored light created when charged particles from solar flares strike the earth's atmosphere near the magnetic poles.

barometer   The gauge used to measure the pressure of the atmosphere.

bartonelosis   A bacterial disease transmitted by sand fleas that causes wartlike sores on the skin in milder cases and anemia in more severe cases.

Big Bang   The theory that the universe we know began its existence about 13 billion years ago as the explosion of a small super-dense concentration of matter.

biodiversity   Describes the number of species in a given environment. It is shorthand for "biological diversity."

biological pump   The biological pump is a natural mechanism whereby carbon from the atmosphere and surface waters of lakes and oceans are cycled into deep waters and sediments. Plants take up the carbon dioxide at the surface of the water. Marine organisms then take up the plants and deposit a fraction of the carbon dioxide on the bottom as waste.

black earth   A type of soil created by early indigenous societies in the Amazon, who mixed charcoal from slowly smoldered trees and plants into the soil. The charcoal then retained plant nutrients, which are usually lost with the tropical rains.

blizzards   A term that was used to describe the dust storms that swept the south central plains of the United States during the Dust Bowl era of the 1930s.

boreal forest   The northernmost type of forest dominated by coniferous trees such as the evergreen spruce, fir, and pine, and the deciduous larch or tamarack.

broadleaf   Trees with broad, flat leaves that are often deciduous—they shed their leaves each year.

carbon burial   The process whereby silicate rocks erode and the chemical breakdown of the rocks converts carbon dioxide to bicarbonate, which is washed to the oceans where it ends up as carbonate sediments at the bottom of the sea.

carbon dioxide   A colorless, odorless gas that passes out of the lungs during respiration. It is also given off by volcanoes. It is the primary greenhouse gas, because it is long-lived (approximately 100 years) and causes the greatest amount of global warming of the long-lived gases.

carbon neutral CO_2 emissions are carbon neutral when they are offset by planting trees, investing in alternative energy, or doing anything else that takes up CO_2 to the point where the offset equals the emission.

carbon sinks   Areas such as forests, grasslands, and the oceans, which naturally take up carbon dioxide from the atmosphere.

celestial sphere   A huge sphere or dome that ancient observers believed surrounded the earth, upon which the stars were fixed. Today, astronomers use the idea as a way to map the location of stars relative to observers on Earth.

cirrus clouds   Clouds that are high and wispy.

clear cutting   Clear cutting is the practice of removing all the timber in a given area, rather than going in and selectively cutting trees. Foresters claim it is necessary for light to rejuvenate seedlings. Environmentalists claim it is a blight, often driven more by economics (it's cheaper to cut down all the trees) than good forestry practices.

climate   The term that refers to the average and common weather conditions of an area over a number of years.

climate change   The natural and manmade phenomena that have made the earth's atmosphere warmer or colder over the ages.

condensation nuclei   The microscopic particles in the air around which cloud droplets and raindrops form.

conifers   Trees with needle-shaped leaves that are evergreen—they stay green all year.

converging currents   Currents in the atmosphere that occur where winds meet or where more air enters an area than departs.

Coriolis effect   The effect caused by the spherical shape of the earth which results in winds veering to the right in the northern hemisphere and to the left in the southern hemisphere.

critical density   That point in the evolution of a star where gas and dust come together close enough for gravity to take over and bind the elements together.

cumulus clouds   Clouds that are puffy and unthreatening; or gathering, rising, and thunderous.

diverging currents   Currents that occur when winds depart in different directions or when more air leaves than arrives.

Dust Bowl   An area in the south central plains of the United States that was devastated by drought in the 1930s.

eccentricity   The shape of the earth's orbit, which varies from round to elliptical.

ecotourism   Environmentally friendly tourism. In North America it refers mostly to minimizing visitor impact. In third-world nations it more often refers to supporting local economic alternatives to depleting natural habitats.

ectotherm   An organism that regulates its body temperature largely by exchanging heat with its surroundings.

endemic   Those species of plants or animals that are found only in restricted areas— a single island, a single river or lake, a single mountain range, or a single mountaintop.

Energy Star   A label on consumer products sanctioned by the U.S. Environmental Protection Agency to help identify energy-efficient products.

ethnobotanists   Scientists who study the cultural and biological relationship of native peoples and the plants in their environment.

frontal cyclones   Large spinning weather systems that occur along turbulent fronts.

frost fairs   Recreational and social events held on the Thames River from the beginning of the 1600s to the beginning of the 1800s during the Little Ice Age, when the river froze over frequently. Since then, the world has gotten warmer, and now the Thames seldom freezes.

gas giants   These are the huge gaseous planets Jupiter, Saturn, Uranus, and Neptune, which have enormous thick atmospheres.

glaciers   Moving masses of ice and snow that accumulate in areas where the rate of snowfall exceeds the rate at which the snow melts.

global warming   The term that refers to recent increases in global temperatures caused by man.

greenhouse effect   This is the consequence that certain gases, chief amongst them carbon dioxide, have on the earth's climate. Though they constitute less than 1 percent of the atmosphere, they act like a blanket covering the earth. Without the greenhouse effect, the average global temperature would be around 0°F instead of the current toasty 59°F.

Holocene   The current or most recent epoch of geologic time, ranging from the present back to the time (about 11,000 years ago) of almost complete withdrawal of the glaciers.

hot air   A term used by environmentalists to apply to greenhouse gas reductions that resulted from weakened economies and the resultant reduction in manufacturing rather than actual efforts put into making those industries operate cleaner and more efficiently.

hurricanes   Violent storms that originate in the tropics and whose winds exceed 73 mph.

hybrid   A vehicle that utilizes both an electrical motor and a gas-powered engine which work in tandem to decrease fuel consumption.

hydroelectric   Power or electricity created by the force of water driving a generator.

ice age   Periods of geological time where Earth was covered by substantial portions of glacial ice. The glaciation of the Pleistocene, which began more than 1.8 million years ago, was not continuous but consisted of several glacial advances interrupted by interglacial stages. Many scientists believe we are currently in such an interglacial stage, which began about 11,000 years ago.

ice sheets   Huge glaciers like those that cover Greenland and Antarctica.

inclination of the axis   A measure of the number of degrees that Earth's axis is tilted away from the plane of its orbit around the sun.

industrial revolution   The period in the economic development of a nation when handicraft production gives way to machine production.

initial conditions   The starting point of a weather forecast that takes into consideration temperature, wind speed and direction, air pressure, and humidity measured from points in a grid overlaying the landscape.

interglacial   The period of relatively stable climate between ice ages. The earth is currently in an interglacial period.

isotopes   The varying forms of an element that have closely related properties but different atomic weights.

jet streams   Strong ribbons of horizontal winds that are found about six to ten miles above the ground in the area between the troposphere, the lower layer of the atmosphere, and the stratosphere above it.

kivas   The ceremonial chambers that surrounded the towns and cliff dwellings of the Anasazi Indians. Their structure was influenced by an earlier day when these Native Americans built homes of covered pits.

mesosphere   The layer of the atmosphere that lies on top of the stratosphere and runs upward to about 50 miles (80 kilometers).

methane   Also called swamp gas, it is the principal component of natural gas. It's given off as part of the decomposition that occurs in wetlands, swamps, and bogs, as well as landfills, rice paddies, and the guts of cattle. It's more efficient at trapping heat than carbon dioxide, but occurs in small concentrations in the atmosphere and has a lifetime of only about 10 to 12 years.

nimbus clouds   Rain-bearing clouds.

obliquity   The term that refers to changes in the tilt of the earth's axis.

old growth forest   The type of forest that has not been previously logged. It represents a more natural and diversified mix of tree and plant species.

ozone   The gaseous, almost colorless form of oxygen that protects the earth against ultraviolet radiation in the upper atmosphere, but is part of the chemical pollution nearer to Earth.

permafrost   The vast stretch of permanently frozen subsoil that stretches across the extreme northern latitudes of North America, Europe, and Asia. This land, too cold for the growth of trees, generally marks the northern limits of the forest.

photosynthesis   The process whereby green plants take in carbon dioxide, mostly through their leaves, and water, mostly through their roots, and mix them together in the presence of sunlight and chlorophyll to produce glucose, a carbohydrate utilized by the plant as energy.

phytoplankton   The microscopic plant life found drifting in the upper sunlit portion of the sea or freshwater, which are the basis of the food chain for other marine organisms.

plate tectonics   Plate tectonics is the theory that the earth is made up of plates that move around, bump, and collide with each other in ways that create some of the earth's most spectacular land forms. The Himalayan Mountains were created when the floating landmass that was India plowed into southern Asia.

precession   The term that refers to the amount of wobble in the earth's axis.

prominences   Huge loops of gases that rise from the surface of the sun and extend far out into space.

proxies   Methods of determining values such as temperatures and rainfall by using substitutes, which give indirect measurements.

relative humidity   The measure of the air's capacity to hold moisture versus the actual moisture present. To get rain you have to have 100 percent humidity.

Renaissance   A term that refers to the great revival of art, literature, and learning that occurred in Europe from the 1300s through the 1500s.

sediments   Fragments of inorganic and organic material that are carried and deposited in layers by wind, water, or ice.

smog   A kind of air pollution that comes in two types. The gray smog of older industrial cities like London and New York derives from the massive combustion of coal and fuel oil in or near the city. The brown smog characteristic of Los Angeles and Denver in the late twentieth century comes from automobiles.

solar flares   Violent eruptions of matter and energy on the sun that spew light and charged atomic particles into space.

storm surges   The tremendous rises in tide that storms like hurricanes push ahead of them.

stratosphere   The layer of the atmosphere just above the troposphere. It rises from 7.5 miles (12 kilometers) to an average of 31 miles (50 kilometers).

stratus clouds   Clouds that are flat and low to the ground.

summer solstice   The time period when the north pole is inclined 23.5 degrees *toward* the sun. In the northern hemisphere this falls on June 21.

systems   From an ecological point of view, systems are niches in the environment that incorporate not just one but various elements such as plants, animals, water, and seasonal weather.

terrestrial planets   The smaller rocky planets Earth, Mars, Mercury, and Venus, which may or may not have thin atmospheres.

thermocline   The boundary between the ocean's warm and cold waters.

Thermohaline Circulation (THC)   A global ocean current that operates like a conveyor belt running through shallow and deep-sea waters to connect the major oceans of the planet.

thermosphere   The uppermost layer of the atmosphere, rising from 50 miles (80 kilometers) to around 400 miles (644 kilometers), though it really has no well-defined upper limit.

tillite   A layer of rock and clay that are evidence of glacial deposits.

troposphere   The bottom layer of the atmosphere, rising from sea level up to an average of about 7.5 miles (12 kilometers).

upwellings   The upward movement of deeper water to the surface.

urban heat island effect   A term used to describe the fact that urban environments are usually warmer than surrounding rural areas.

weather   The current conditions of temperature, humidity, wind, and precipitation.

winter solstice   The time period when the north pole is inclined 23.5 degrees *away* from the sun. In the northern hemisphere this falls on December 21.

Younger Dryas event   The event occurred about 12,800 years ago during a period of increasing warmth, when the world was suddenly and very drastically thrown back into the last glacial period of the ice age. It remained that way until about 11,500 years ago when it swung back out.

Further Reading

Books on Climate and Climate Change

Alley, Richard B. *The Two-Mile Time Machine: Ice Cores, Abrupt Climate Change, and Our Future*. Princeton University Press, 2000. Assessment Reports of the Intergovernmental Panel on Climate Change ("IPCC").

Casten, Thomas R. *Turning Off the Heat*. Prometheus Books, 1998.

Climate Change 2007: The Physical Science Basis. Working Group I. Cambridge University Press, 2007.

Climate Change 2007: Impacts, Adaptation, and Vulnerability: Working Group II. Cambridge University Press, 2007.

Climate Change 2007: Mitigation of Climate Change: Working Group III. Cambridge University Press, 2007. Committee on the Science of Climate Change, National Research Council.

Climate Change Science: An Analysis of Some Key Questions. National Academy Press, 2001.

Gershon, David. *Low-Carbon Diet: A 30-Day Program to Lose 5,000 Pounds— Be Part of the Global Warming Solution!* Empowerment Institute, 2006.

Gore, Al. *An Inconvenient Truth: The Planetary Emergency of Global Warming and What We Can Do About It.* Rodale, 2006.

Goldstein, Dr. Mel. *The Complete Idiot's Guide to Weather.* Alpha Books, 1999.

Harris, Paul G., ed. *Climate Change and American Foreign Policy.* Hampshire, United Kingdom: Palgrave, 2000.

Houghton, John T. *Global Warming: The Complete Briefing.* Cambridge University Press, 1997.

Kolbert, Elizabeth. *Field Notes from a Catastrophe: Man, Nature, and Climate Change.* Bloomsbury, 2006.

Lamb, H. H. *Climate, History, and the Modern World.* Routledge, 1997.

Lomborg, Bjorn. *Cool It: The Skeptical Environmentalist's Guide to Global Warming.* Cyan-Marshall Cavendish, 2007.

Lutgens, Frederick K. *The Atmosphere: An Introduction to Meteorology.* Prentice Hall, 1989.

National Research Council. *Abrupt Climate Change: Inevitable Surprises.* National Academy Press, 2002.

Root, Terry, and Stephen H. Schneider, ed. *Wildlife Responses to Climate Changes.* Island Press, 2002.

Somerville, Richard C. J. *The Forgiving Air: Understanding Environmental Change.* University of California Press, 1996.

Field Guides

Chartrand, Mark R. *National Audubon Society Field Guide to the Night Sky.* Knopf and Chanticleer Press, 1997.

Day, John A., and Vincent J. Schaefer. *A Field Guide to the Atmosphere.* Houghton Mifflin, 1999.

Ludlum, David M. *National Audubon Society Field Guide to North American Weather.* Knopf, 1991.

Books on the Environment

Berger, John J. *Charging Ahead: The Business of Renewable Energy and What It Means for America*. University of California Press, 1998.

Burton, John, ed. *The Atlas of Endangered Species*. Macmillan, 1998.

Ehrlich, Paul R., and Anne H. Ehrlich. *One with Nineveh: Politics, Consumption, and the Human Future*. Island Press, 2004.

Gay, Kathlyn. *Rainforests of the World*. ABC-CLIO, Inc., 2001.

Honey, Martha. *Ecotourism and Sustainable Development: Who Owns Paradise?* Island Press, 1998.

Hurst, Phillip. *Rainforest Politics: Ecological Destruction in Southeast Asia*. Humanities Press International, 1990.

Jackson, Rob. *The Earth Remains Forever: Generations at a Crossroads*. University of Texas Press, 2002.

Myers, Norman. *The Primary Source: Tropical Forests and Our Future*. W. W. Norton & Company, 1992.

Revkin, Andrew. *The Burning Season: The Murder of Chico Mendes and the Fight for the Amazon Rainforest*. Houghton Mifflin, 1990.

Wilson, Edward O. *The Diversity of Life*. Harvard University Press, 1992.

Concerned Organizations

Center for Conservation Biology (CCB)
Department of Biological Sciences
Stanford University
Stanford, CA 94305-5020
Phone: 650-723-3171
Fax: 650-723-5920
www.stanford.edu/group/CCB/

Conservation International (CI)
2011 Crystal Drive, Suite 500
Arlington, VA 22202
Phone: 703-341-2400
Toll-free: 1-800-429-5660
www.conservation.org

Conservation Fund
1655 N. Fort Myer Drive
Suite 1300
Arlington, VA 22209-3199
Phone: 703-525-6300
Fax: 703-525-4610
www.conservationfund.org

Environmental Defense (ED)
257 Park Avenue South
New York, NY 10010
Phone: 212-505-2100
Toll-free: 1-800-684-3322
Fax: 212-505-2375
www.environmentaldefense.org

Greenpeace USA
702 H Street NW
Washington, DC 20001
Phone: 202-462-1177
Toll-free: 1-800-326-0959
www.greenpeaceusa.org

National Audubon Society
700 Broadway
New York, NY 10003
Phone: 212-979-3000
Fax: 212-979-318
www.audubon.org

National Wildlife Federation (NWF)
11100 Wildlife Center Drive
Reston, VA 20190-5362
Phone: 1-800-822-9919
www.nwf.org

Native Energy
823 Ferry Road
P.O. Box 539
Charlotte, VT 05445
Phone: 1-800-924-6826
www.nativeenergy.com

Natural Resources Defense Council (NRDC)
40 West Twentieth Street
New York, NY 10011
Phone: 212-727-2700
Fax: 212-727-1773
www.nrdc.org

The Nature Conservatory
4245 N. Fairfax Drive, Suite 100
Arlington, VA 22203-1606
Phone: 1-800-628-6860
www.nature.org

Rainforest Alliance
665 Broadway, Suite 500
New York, NY 10012
Phone: 212-677-1900
Toll-free: 1-888-MY-EARTH
www.rainforest-alliance.org

Sierra Club
85 Second Street, Second Floor
San Francisco, CA 94105
Phone: 415-977-5500
Fax: 415-977-5799
www.sierraclub.org

Smithsonian Institution (SI)
P.O. Box 37012
SI Building, Room 153, MRC 010
Washington, D.C. 20013-7012
Washington, D.C. 20560
Phone: 202-633-1000
www.si.edu

Survival International
6 Charterhouse Buildings
London EC1M 7ET
United Kingdom
Phone: 44 20 7687 8700
Fax: 44 20 7687 8701
www.survival-international.org

Union of Concerned Scientists (UCS)
2 Brattle Square
Cambridge, MA 02238-9105
Phone: 617-547-5552
Fax: 617-864-9405
www.ucsusa.org

United Nations Environment Programme (UNEP)
United Nations Avenue, Gigiri
P.O. Box 30552, 00100
Nairobi, Kenya
Phone: (254-20) 7621234
Fax: (254-20) 7624489/9
ww.unep.org

The Wilderness Society
1615 M Street NW
Washington, D.C. 20036
Phone: 1-800-843-9453
www.wilderness.org

World Conservation Union (IUCN)
Rue Mauverney 28
Gland
1196
Switzerland
Director General's Office
Phone: +41 (22) 999-0000
Fax: +41 (22) 999-0002
www.iucn.org

World Wildlife Fund (WWF)
Avenue du Mont-Blanc 1196
Gland
Switzerland
Phone: 41-22-364-91-11
Fax: 41-22-364-88-36
www.panda.org

Index

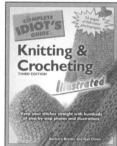